PONDEROSA

People, Fire, and the West's Most Iconic Tree

Carl E. Fiedler

and

Stephen F. Arno

2015
Mountain Press Publishing Company
Missoula, Montana

Library of Congress Cataloging-in-Publication Data

Fiedler, Carl E.
 Ponderosa : people, fire, and the West's most iconic tree / Carl E. Fiedler
and Stephen F. Arno.
     pages cm
 Includes bibliographical references and index.
 ISBN 978-0-87842-638-6 (pbk. : alk. paper)
 1. Ponderosa pine. 2. Forests and forestry—Fire management—West (U.S.)
 3. Forest restoration—West (U.S.) I. Arno, Stephen F. II. Title.
 SD397.P6115F52 2015
 634.9′6180978—dc23
                                2015004960

Printed in Hong Kong by Mantec Production Company

MP Mountain Press
PUBLISHING COMPANY
P.O. Box 2399 · Missoula, MT 59806 · 406-728-1900
800-234-5308 · info@mtnpress.com
www.mountain-press.com

CONTENTS

Ponderosa pines occasionally grow in tight clumps. —*Photo by Lance Schelvan*

PREFACE

I have put all I had to say into the body of this book; but, being informed that a preface is a necessary evil, I have written this one.

—Dan DeQuille, *History of the Big Bonanza*, 1876

I'VE OFTEN BEEN ASKED IN THE LAST FEW YEARS, "What have you been doing?" When I answer, "Writing a book on ponderosa pine," the typical response is some variation of the question "How can you write a book about a tree?" The implication is that how in the world would there be enough material to write such a book. My response is that it is indeed difficult. But I do not elaborate why, because it is for a different reason than one might think. It is not how difficult it is to find enough material, but rather how to fit the wealth of knowledge and stories about ponderosa pine's place in western life between two covers.

Since the mid-1980s, I have had the privilege of conducting workshops on ponderosa pine management and restoration forestry for public agencies and Indian tribes in nearly every western state. These experiences, plus visits to all but a few of the ponderosa pine places profiled in part II of this book, opened my eyes to an ever larger and more fascinating world of ponderosa pines and often left me in awe. Our aim is that *Ponderosa: People, Fire, and the West's Most Iconic Tree* captures and communicates this passion to the reader.

The initial intrigue of writing a book about my favorite tree—ponderosa pine—soon became a sobering reality. How does one go about weaving together the technical and the beautiful, the ecological and the utilitarian? I recall an experience I had while conducting a training session on ponderosa pine management in New Mexico some years ago that illustrates this dichotomy. I had asked the late Wendell Chino, charismatic tribal president of the Mescalero Apache, if he would give a short welcome to the class. Chino agreed and told a story about going on a field trip one day with his forestry staff to view ongoing projects on the reservation. At the start of the trip, Mr. Chino noticed an especially tall, graceful old pine. After admiring it for a minute, he wondered out loud how tall it might be. One of the foresters overheard and responded that he would measure it and let Mr. Chino know. When the group returned to the parking area, Chino was shocked to see the large, old pine lying on the ground. The forester standing next to the prone tree proudly volunteered, "It's 110 feet long, Mr. Chino." Chino was quiet for a minute and then responded coldly, "I didn't ask how long it was, I asked how tall it was."

I also recalled Mark Twain's thoughts on how to organize and write a book: "Ideally a book would have no order to it, and the reader would have to discover his own."[1] How tempting it was to heed Twain's words. But there was another option—pictures. While this is not a picture book, we soon found that it is a book that we could not do without pictures. As John Wesley Powell wrote upon returning from his exploration of the American West, "Realizing the difficulty of painting in word colors a land so strange, so wonderful, and so vast in its features, in the weakness of my descriptive powers I have sought refuge in graphic illustration."[2] We, too, seek refuge in pictures, knowing full well that pictures also fall short of the wonder of personally experiencing ponderosa pine forests.

In the best Twain tradition, then, this book is broadly aimed—at students, hikers, hunters, ecologists, loggers, forest landowners, forest dwellers, tourists, nature lovers, Sunday afternoon drivers, and ponderosa pine aficionados—wherever they may be.

Donald Worster, who reviewed Patricia Limerick's western classic *The Legacy of Conquest*, posed a rhetorical question that gave us pause: "Is it a book that will open a door in a cul-de-sac?"[3] That is the real challenge in writing a book about a tree. In an era of social media, rapidly changing technology, and instant gratification, can a book about ponderosa pine compete for people's time and attention?

In the end, Steve Arno and I could not keep from writing this book. After spending our careers studying, working, and traveling in ponderosa pine forests, we were hooked. Perhaps a quote from Norman Maclean's acclaimed *A River Runs Through It* says it best: "On the Big Blackfoot River above the mouth of Belmont Creek the banks are fringed by large ponderosa pines. In the slanting sun of late afternoon the shadows of great branches reached from across the river and the trees took the river in their arms. The shadows continued up the bank, until they included us."[4]

—Carl E. Fiedler

MY GRANDFATHER WAS BORN IN 1865 AT SUSANVILLE, California, where the rough wagon road entered a virgin ponderosa pine forest that covered the broad northern end of the Sierra Nevada. This was already sixteen years after the California Gold Rush had lured tens of thousands of would-be prospectors and adventurers into the Sierra's ponderosa pine forest. These forty-niners soon displaced thousands of Indians who had dwelt in the forest since time immemorial. The newcomers brought superior technology for building a home in the woods—long crosscut saws for felling and bucking the big trees into stove-wood and logs, and sawmills for slicing them into lumber.

In the 1920s when Aunt Ruth and her husband settled on a place in those ponderosa forests, up the North Yuba River, they still wielded crosscut saws, known as misery whips. They also continued to heat and cook with woodstoves. They could travel in Model Ts and other early motorcars, but they still needed horses to pull cars out of mud holes.

During the summers of 1963 and 1965 I lived in the Sierra Nevada's splendid ponderosa pine–mixed conifer forest while working as a ranger and then a naturalist in Sequoia and Kings Canyon National Parks. Age-old, black fire scars on sequoias and ponderosa pines testified to a long history of low-intensity burning, and the Park Service was considering reintroduction of fire to restore more open, natural conditions. I led evening campfire programs in a primitive outdoor amphitheater and surprised many park visitors when I started the campfire by simply igniting an armload of dry pine needles and cones.

At summer's end in 1965, my wife and I moved into ponderosa pine country in the Northern Rockies, where I took up graduate studies in forest ecology at the University of Montana in Missoula. By then most of the remote homesites in the woods had been abandoned, people having moved to town for jobs and conveniences like electricity. However, in a few years that trend reversed and thousands of families, including ours, bought a parcel of ponderosa pine forest and moved into the woods. The back-to-the-land movement was exemplified by the first Earth Day in 1970. This cultural change was aided by advancing technology—comfortable four-wheel-drive vehicles with optional snowplows, expanding power and phone networks, and increasingly available mortgage loans for remote homesites.

My background in forestry made obvious the need to thin out our seriously overcrowded, second-growth ponderosa pine forest before we could consider building a home there. In 1972 when the remaining old-growth ponderosas were logged immediately adjacent to our property, I noticed that the fresh stumps displayed cross-sections of repeated scars from surface fires that dated (by tree-ring counts) from the early 1600s until the turn of the twentieth century. This pattern of frequent, low-intensity fires was largely unknown in the Northern Rockies, and as a new US Forest Service scientist, I had the opportunity over the next several years to study fire history in many locations in different forest types.

In the meantime, in 1973 we began thinning our 60-acre family forest, with chainsaw and farm tractor, and burning the branches and treetops. As we established a home on this property, we decided to help restore a generally open-grown forest featuring big old pines but having enough younger trees to perpetuate it. By the mid-1980s my research and that of many other ecologists showed that this variation in tree age had originally characterized ponderosa pine forests throughout much of the West.

By the late 1980s unprecedented, severe wildfires were occurring in many ponderosa forests throughout the West. I began collaborating with fellow Forest Service researcher Mick Harrington and University of Montana research professor Carl Fiedler, who had been conducting prescribed fire and silviculture cutting treatments to help re-create ponderosa forests that would be resistant to major damage from wildfire and outbreaks of insects and disease. We and other colleagues published many technical articles to inform foresters and other specialists about the ecology of ponderosa pine forests and implications for forest management. However, millions of Americans have a stake in the management and perpetuation of these backyard and nearby forests, and so we have aimed this book at telling the ponderosa pine story to all of these important stakeholders.

—Stephen F. Arno

ACKNOWLEDGMENTS

FIRST AND FOREMOST, WE THANK OUR PUBLISHER, Mountain Press, for sustained interest and support in bringing this rather unusual book to fruition. Jennifer Carey, Mountain Press editor, provided a keen eye and invaluable guidance in organizing chapters and tightening our focus. We are especially grateful to Jim Habeck, retired ecology professor at the University of Montana, and consulting forester Matt Arno (Stephen's son), for reviewing the entire manuscript. Jim has lifetime interest in fire-dependent forests, and Matt's concern for ponderosa restoration runs deep. We also acknowledge Jerry Williams, retired Forest Service national director of fire management, Michael "Mick" Harrington, retired Forest Service research forester, and Dr. Mark Finney, Forest Service fire scientist, for reviewing substantial portions of the manuscript. The following people reviewed chapters or provided important topical information: Chapter 2: Indians in the Pines (Lawrence Kingsbury, Mark Petruncio, Jim Roessler), Chapter 3: Pioneers in the Pines (Dabney Ford, Paul Horsted, Jim Roessler), Chapter 8: Logging Legacy—From Clearfelling to Clearcutting (Rolan Becker, Richy Harrod, Marlin Johnson, Sonny LaSalle, Dave Powell, H. B. "Doc" Smith), Chapter 9: Loving the Forests to Death (John Lehmkuhl), Chapter 10: Forests under Siege—From Megafires to Bark Beetles (Bill Armstrong, Stephanie Coleman, Chris Fettig, Jim Paxon, Gayle Richardson, Rory Steinke, Diane Vosick, Jim Youtz), Chapter 11: Restoration—Is It Too Late? (Mike Anderson, Jackie Banks, Craig Bienz, Anne Bradley, Dick Fleishman, Paul Harlan, Mary Lata, Chuck Lewis, Amanda McAdams, Neil McCusker, Jane O'Keeffe, Chris Pileski, Henry Provencio, Carl Skinner, Diane Vosick, Jim Walls, Amy Waltz, Mike Williams, Jim Youtz).

Part II of our book features ponderosa pine places in every western state. The following people either suggested sites or provided useful supporting information: Arizona (Chris Baisan, Wally Covington, Don Falk, Calvin Farris, Pete Fulé, Jim Malusa, Suzanne Moody), California (Bob Means, Carl Skinner, Detlev Vogler), Colorado (Peter Brown, Mark Krabath, Laurie Swisher), Idaho (Chad Hood), Montana (Emily Guiberson, Bob Means, Dennis Sandbak), Nebraska (Sybil Malmberg Berndt, Richard Gilbert), Nevada (David Charlet, Martha Williamson), New Mexico (Anne Bradley, Don Falk), Oklahoma (Monty Joe Roberts), Oregon (Stephen Fitzgerald,

Dave Powell), South Dakota (Peter Brown), Texas (Edna Flores, Jason Wrinkle), Utah (Don Hanley, Michael Kuhns, Darren McAvoy, Morgan Mendenhall, Doug Page), Washington (Kyle Dodson, Kevin Zobrist), Wyoming (Bob Means), British Columbia (Bob Gray).

We are grateful to those who provided translations, historical information, or photographs from museum, library, or archive collections. Notable help came from Lee Brumbaugh, Nevada Historical Museum; Sean Evans and Jess Vogelsang, Cline Library, and Diane Vosick, Ecological Restoration Institute, Northern Arizona University; Mark Fritch, Mansfield Library, University of Montana; George Gruell, Carson City, Nevada; Ted Hughes, Missoula Art Museum; Vanessa Ivey, Des Chutes Historical Museum; Eben Lehman, Forest History Society; Marker Marshall, Grand Canyon National Park; Anne Martinez and Roger Myers, University Libraries, University of Arizona; Crystal Miles, Bancroft Library, University of California, Berkeley; Keith Moser, US Forest Service Rocky Mountain Research Station; Darrell Mullins, Tehama County Museum; Nic Munagian, Charles Deering McCormick Library, Northwestern University; Laurie Porth, US Forest Service western regional archivist; Angelica Sanchez-Clark, National Park Service/University of New Mexico; Christine Stokes, Jay Thompson, Jeremy Tuggle, and volunteer Joann Montgomery, Shasta Historical Society; George Thompson, Meriam Library, Cal State University, Chico; Rose Trujillo, Chaco Culture National Historical Park; and Bill Whitfield, Ravalli County Museum.

Photographers who allowed us to use their images greatly enhanced our ability to tell the ponderosa pine story. These individuals are identified with their respective photographs at appropriate places throughout the book. Finally, we acknowledge Peter Brown, Jim Habeck, Bob Means, Doug Page, and Diane Vosick for their sustained interest and help throughout the project.

We thank you all.

PART I

View across ponderosa pine savanna from Upper Beaver Meadows loop trail, Rocky Mountain National Park, Colorado.

A Tale of Two Forests

If you know your West at all, you know its Yellow Pine
[which is found] in every western state . . . and grows
most abundantly in the West's prime "Vacationland." . . .
Its dry and spacious groves invite you to camp among
them. . . . Its great boles and boughs frame many of the
grandest views.

—Donald Culross Peattie, *A Natural History of Western Trees*, 1950

BONANZA, THE TOP-RATED TELEVISION EPIC of the 1960s, featured the Ponderosa, a historic western ranch named for its majestic ponderosa pines. As the first hour-long TV series filmed in color, *Bonanza* captivated viewers with images of parklike ponderosa forests—a landscape that also inspired many early western explorers and settlers. After struggling for months to cross scorching plains and deserts, wagon train pioneers rejoiced at the first sight of tall trees as they approached Oregon and California. These travelers soon entered a realm of giant pines and grassy glades—well-spaced trees with spreading limbs and towering trunks clad with platy, cinnamon-colored bark.

In 1853 on the Oregon Trail, Rebecca Ketcham recorded her impressions of these forests from the hard seat of a covered wagon—or likely, while walking to lessen the load on worn-out oxen: "Our road has been nearly the whole day through the woods, that is, if beautiful groves of [ponderosa] pine trees can be called woods. . . . The country all through is burnt over, so often there is not the least underbrush, but the grass grows thick and beautiful."[1] Ms. Ketcham's journal account vividly described the effects of frequent low-intensity fires that historically burned along the ground, keeping the forest open and inviting. Seven hundred miles to the south, a young

3

Army Lieutenant was similarly impressed. Riding through the unexplored forests of northern Arizona in 1857, Lt. Edward Beale summed up his view from the saddle: "It is the most beautiful region. . . . A vast forest of gigantic [ponderosa] pines, intersected frequently with open glades . . . and covered with the richest grasses, was traversed by our party for many days."[2] Little wonder that modern Americans also are captivated by these pines, the only tall trees that break the monotony of a sprawling, parched landscape.

The remarkably drought-tolerant ponderosa pine forms forests that border arid grasslands, shrublands, and bushy juniper woodlands in every western state. The widespread use of ponderosa's name on everything from towns and businesses to highways, schools, and cemeteries testifies to its integral place in everyday life and its stature as a symbol of the West. In 1908, school children in Helena, Montana, voted ponderosa pine the tree that best represented their state, and many years later the legislature designated it as the state tree. Ponderosas are simply the most widespread and recognized trees in the West, icons of our backyards.

Ponderosa pine graces the sunny interior valleys of southern British Columbia and extends southward at increasing elevations to the mountain recreation areas of southern California. While not inhabiting the humid Pacific coastal strip, ponderosa forests are sprinkled across the semiarid

A wagon passes through an open ponderosa forest on the Tusayan National Forest, Arizona, in 1909. —*Photo by G. A. Pearson, US Forest Service*

West, extending 1,000 miles inland to South Dakota's Black Hills, the Sand Hills of Nebraska, and mountains jutting high above the deserts in southern New Mexico, West Texas, and northern Mexico. This splendid pine prospers under conditions that elsewhere in the world produce only brushlands or dwarf trees.

The prominent western cities of Spokane, Coeur d'Alene, Bend, and Flagstaff are built entirely within ponderosa forests, as are numerous smaller cities and towns and major recreation areas, such as Lake Tahoe, southern California's Big Bear Lake, and South Dakota's Black Hills. Suburbs and out-lying areas of Boise, Missoula, Helena, and Rapid City also sit within the pleasing ponderosa pine zone, as do the foothills adjacent to Fort Collins, Boulder, Denver, Colorado Springs, and Santa Fe. Each year more and more houses are built in these residential forests, often adjacent to national forests and other public land. Known as the wildland urban interface (WUI), the number of homes in the woods has burgeoned. During the 1990s alone, 2.2 million housing units were added to the WUI in the West, many of them in ponderosa pine forests.[3]

The rapidly growing WUI now represents a huge problem. The famil-iar and cherished ponderosa forest has been heavily impacted by humans and is now a deteriorating and hazardous vestige of what the early settlers found. Since the 1980s ponderosa pine forests have become the epicenter of disastrous western wildfires that often consume hundreds of homes during a single fire season.

There's a reason that vast numbers of homes are now located in pon-derosa forests. Ponderosa is the most widespread tree in the western United States, and it occupies a warm, dry forest habitat within or close to major western valleys. Ponderosa forests are places of moderation—neither too hot nor too cold—and are some of the most desirable places to live in the West. Native Americans lived among the pines for untold centuries, and now mod-ern Americans do too.

It is not the historical forest of giant pines and grassy glades that is fuel-ing the catastrophic fires that consume the houses of today. Fires burned through the historical forest frequently but caused little damage. The modern forest is different, still replete with ponderosa pine but changed by a century of timber harvest and fire suppression. Well over 90 percent of the original ponderosa pine forests were logged heavily, often more than once. Most of the characteristic big trees, which were hundreds of years old, are long gone. Overcrowded, second-growth stands typify the modern forest, and on perhaps half of the land originally dominated by large, old ponderosas, firs are replacing the younger pines.

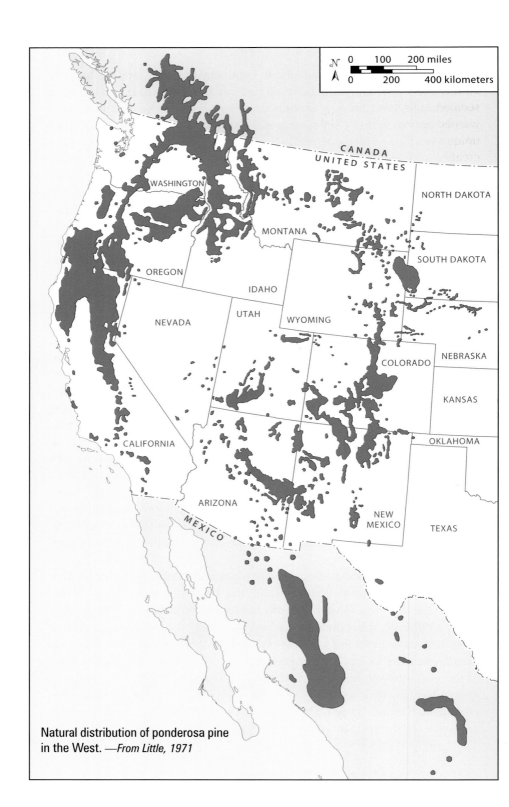

Natural distribution of ponderosa pine
in the West. —*From Little, 1971*

In the early 1900s, the US Forest Service led a crusade to extinguish all fire in the forest to protect timber, watersheds, and people, a policy that seemed entirely reasonable at the time. However, even then some people warned against trying to completely eliminate the fires that had shaped these unique forests for countless centuries. Nature's free-roaming fires helped create open ponderosa forests with well-spaced, majestic trees, but fire's role in this relationship didn't become known until late in the twentieth century, as the science of ecology matured. Because fires have been suppressed for 100 years, the forest canopy just gets thicker and pine needles and other fuels continue to accumulate. Now, when fires start, their unprecedented severity can kill nearly all trees, degrade or destroy endangered species habitat, damage forest soils, and trigger erosion and stream sedimentation. The overcrowded forests are also stressed and susceptible to bark beetles and disease. As Steve Arno and his coauthor put it in a 2002 book, "Simply leaving today's forests alone after a century of fire suppression and logging of big trees is not caring for them; it is abandonment."[4]

These dense forests do provide secluded homesites, but at a cost. Hundreds of millions of dollars and even some firefighter lives are expended trying to save houses from today's raging wildfires, but if they are saved, the houses stand amidst a charred wasteland of dead trees. Recognizing the threat, more forest landowners and federal land managers are now thinning and using prescribed burning to mimic the low-intensity fires that historically reduced forest fuels, killed small trees, and sustained magnificent ponderosa pine forests. Such efforts are vital to the health of these forests, but millions of acres must still be treated.

This book, then, chronicles the history, ecology, and allure of the original ponderosa forest in western North America and its importance in the everyday life of Native Americans, explorers, and early settlers. It also profiles the century-long transformation to the modern forest. Human-caused changes in historical ponderosa pine forests unintentionally transformed their durability into vulnerability. Understanding the character of yesterday's ponderosa forests provides insights into how they might be restored today. Our book also explains the science behind restoring and sustaining these forests into the future, outlines how to make both a forest home and its surrounding forest fire resistant, and guides the reader toward some memorable ponderosa trees and forests in each western state and British Columbia.

INDIANS IN THE PINES

This species [ponderosa pine] also gives forth the finest music to the wind.

—John Muir, *The Mountains of California*, 1907

THE FIRST AMERICANS BEGAN CROSSING over the Bering Strait land bridge from Asia into North America about 13,000 years ago as the last ice age came to an end. They gradually moved south along ice-free corridors, first encountering ponderosa pine in present-day California or the Southwest, because it didn't grow as far north during the glacial period as it does today. The tree's oldest documented use in the Southwest was for corner posts in partially underground pit houses, which were a primary form of shelter until about AD 500.[1] Pine logs were sometimes used for entryways and wall supports in these structures, or to shore up roofing materials.

The next step forward in the early use of ponderosa pine for home building was a giant one indeed. The Anasazi used ponderosa timbers in the construction of their magnificent pueblo complex in Chaco Canyon in northwest New Mexico, beginning about AD 850.[2] Anasazi buildings were multistoried, multipurpose community structures—impressive in size, form, and function—not just in the Chacoan Pueblo era but in any era. The most acclaimed of these is Pueblo Bonito, which means "pretty town" in Spanish. It is also the largest of the twelve great houses in the Chaco Canyon complex. This sandstone, mortar, and pine-timbered structure consisted of about 650 rooms, including food storage areas, residential living areas, and over 30 circular, multistoried kivas used in ceremonies.

Construction of the Anasazi pueblo complex required staggering amounts of building materials, including an estimated 200,000 trees[3]—most of them ponderosa pine. Given that the dry forest sites in the surrounding mountains support relatively few ponderosa pines per acre of the size used in pueblo construction, large areas must have been harvested to provide the necessary material.

Even more astonishing, the closest ponderosa pine forests lie 50 miles to the southwest in the Chuska Mountains. How hundreds of thousands of ponderosa logs were transported such a long distance, with no apparent physical damage, remains an enigma. Neither horses nor other beasts of burden were available at the time. One theory is that the logs were elevated above the ground in some way, possibly transported in slings or on workers' shoulders, rather than being dragged, which would help explain the lack of scarring or compression damage to the timbers. Regardless of the method of transport, it is a monumental feat that rivals construction of the pyramids of Egypt. Scientists have recently employed sophisticated strontium isotope analysis to compare timbers in the pueblos with pine trees in various surrounding mountain ranges. Results confirm that some of the ponderosa pine timbers used to construct Pueblo Bonito did come from the Chuska Mountains, but many others came from even more distant sources, including the La Plata and San Juan Mountains, located more than 100 miles to the northwest in Colorado.[4]

Ponderosa pine timbers protruding from the ruins of Pueblo Bonito, Chaco Culture National Historical Park, New Mexico. —*Photo courtesy National Park Service*

The ponderosa logs were used primarily as timber supports, lintels (beams above doorways), and vigas (rafters or roof supports). Logs were carefully finished before use. Some pine roof supports that have remained intact over the centuries feature perpendicular cuts so smooth that it appears they were made with a crosscut saw, even though such saws did not exist at the time.[5]

The remains of the Anasazi pueblo complex were first discovered by US Army Lt. James Simpson in 1849 while on a military expedition. Expedition members briefly examined the ruins, and Simpson described their findings in his trip report.[6] However, comprehensive study of the ruins would not begin until 1896, when a rancher-archaeologist and a representative from the American Museum of Natural History began excavations at Pueblo Bonito.[7] After the initial excitement of collecting artifacts, uncovering nearly two hundred rooms, and mapping the structures within the Chaco complex, one nagging question remained: When did Anasazi construction of this archaeological treasure occur? The key to cracking the code came from a most unlikely source using an equally unlikely material—ponderosa pine.

In the early 1900s, an astronomer from Arizona named Andrew Douglass began to investigate the use of tree rings as a long-term record of sunspot activity. He examined old ponderosa trees and stumps near Flagstaff and discovered that tree rings reflect high- or low-moisture years in the form of wide or narrow annual growth rings. He found that there was a high degree of similarity in the ring-width patterns among different trees. Comparing the widths and patterns of overlapping tree-ring sequences allowed him to eliminate problems with occasional false rings or rings that were otherwise hard to interpret—a technique called cross dating. In 1911, he discovered that ponderosas near Prescott, Arizona, shared similar ring patterns with those around Flagstaff, some 60 miles away. But it was an archaeologist named Clark Wissler who first considered the potential application of tree rings for dating archaeological ruins.

In the spring of 1914, Wissler wrote Douglass after reading Douglass's paper, "A Method for Estimating Rainfall by the Growth of Trees," and raised the possibility of using tree-ring analysis to assign dates to archaeological wood samples.[8] Thus began a fifteen-year quest to establish a continuous tree-ring chronology using ring-width patterns in the ponderosa timbers for dating construction of the pueblos. Douglass was able to develop a 450-year-long chronology using living ponderosa pines in the Flagstaff vicinity. He was also able to build an even longer undated, or floating, chronology based on ponderosa timbers from the pueblo ruins. What was missing was a cross-dated series that bridged the gap between the modern and prehistoric tree-ring chronologies. Fortuitously, archaeologists working with Douglass in the summer of 1929 found just such a missing link, a charred ponderosa timber in the Whipple Ruin near Show Low, Arizona.[9] Archaeologists labeled this specimen HH-39, but Douglass likened it to the Rosetta stone, which provided the key to translating Egyptian hieroglyphics. HH-39 completed a sequence back to AD 701 that allowed him to accurately and precisely date forty other archaeological sites across the American Southwest.[10] Then, based on the new chronologies added from those sites, he determined that construction at Pueblo Bonito began in AD 919, although more recent research has identified even older timbers and moved the start date back to about AD 850.[11] Douglass's novel approach also provided a regional framework for establishing other important dates in the development and decline of pueblos throughout the Southwest.

The major pueblo complexes were rapidly abandoned starting about AD 1130. This date coincides with the beginning of a fifty-year drought as deciphered from Douglass's 1,250-year-long ponderosa pine tree-ring chronology, but abandonment was likely due to other factors as well.[12] Pueblo residents, who were primarily agriculturists, dispersed to other settlements or started new ones.

Legendary tree-ring chronologist Andrew E. Douglass examining a pencil-shaped tree core for annual growth rings. The sample is obtained by drilling into the center of a tree with a tool called an increment borer and extracting the wood core. —*Photo courtesy Special Collections at the University of Arizona Libraries and the University of Arizona Laboratory of Tree-Ring Research*

An Indian camp in an open ponderosa pine forest in central Washington. The Yakama Indians came here to dig camas roots during the spring. —*Photo by Edward S. Curtis, 1910, courtesy Charles Deering McCormick Library of Special Collections, Northwestern University Library*

Nomadic hunter-gatherers elsewhere in the West also developed a close relationship with ponderosa pine. Because ponderosa forests were places of climatic moderation, they were desirable places to live for humans and animals. Pine forests, which share boundaries with drier grasslands below and moister, mixed conifer forests above, were prime hunting areas because they provided animals a range of food sources and adjacent cover.

Ponderosa pine, being the most abundant and ubiquitous forest tree in the desirable areas to live, was a staple of Indian life. In addition to protection from sun and wind and the presence of large limbs to suspend food out of the reach of animals, ponderosas contributed to all of life's basic needs, including fuel for heat, timber for shelter, bark and pine seeds for food, and pitch for sealant and glue. For no tribe was this more true than for the Hualapai, or Walapai, who still live in a 100-mile-wide strip of pine forest and grassland south of the Colorado River in Arizona. The tribe's name is derived from *hwa:l*, their word for ponderosa pine; hence the term *Hualapai*, meaning "people of the tall pine." [13]

Native Americans harvested the inner bark of ponderosa pines for food by peeling the bark of mature trees. For some tribes, peeling ponderosa pine bark appears to have been an annual tradition, much like berry picking, digging roots, or salmon fishing. For others, bark was a food of last resort before starvation. Bark was typically peeled over a several week period in the spring, a time when the sap was flowing and the inner bark was at its sweetest and most nutritious. Women, helped by their children, would select the trees for peeling and sample the bark as spring progressed until the bark became loose and slippery, making the still-difficult task of peeling a bit easier.

The peeler first cut two horizontal slits in the bark, one a foot or two above the ground and the other above her head. Next came a vertical cut connecting the two horizontal ones, which allowed for insertion of a long, sharpened, fire-hardened stick used like a giant chisel to pry the bark loose. The peeled area was typically 1 foot or so wide and 4 to 8 feet high, but not so large as to kill the tree. The sticky inner bark, or phloem layer, was then removed in strips or separated using a scraping tool. The inner bark, which has the consistency of saturated felt, was often consumed fresh off the tree as chewy candy. Strips of inner bark were sometimes boiled with meat to make a stew, or slow cooked over rocks to improve digestibility. Most often, the peeled bark was ground into flour for use throughout the year, when it might be baked into cakes or mixed with dried meat or berries.

A park ranger sampling the vanilla aroma of a bark-peeled tree in the Indian Grove, Great Sand Dunes National Park, Colorado. —*Photo by Patrick Myers, National Park Service*

Chemical analyses have identified the high nutritional value of pine bark components. For example, one pound of peeled inner bark has the calcium equivalent of nine glasses of milk.[14] Pine bark also contains high concentrations of antioxidants and is a good source of fiber, protein, carbohydrates, vitamin C, iron, and trace minerals, especially at the beginning of active growth in the spring. Bark helped balance Indian diets that were otherwise mainly protein from fall through spring. However, harvesting bark required a considerable expenditure of energy, especially if it was later ground into flour, and a scientist studying native foods in the early 1900s reported that it was hard to digest.[15]

Bark-peeling by native people was likely a more important and widespread phenomenon than studies of individual bark-peel sites and piecemeal reports would suggest. Wyoming is the only state west of the Great Plains with no documented reports of bark-peeled ponderosa pines. A comprehensive overview of the many reports and studies of ponderosa bark-peeling suggests that Indians in the inland Northwest were the most consistent users of bark for food. For example, the Okanagan-Colville tribe held ceremonies every spring to celebrate the upcoming bark harvest, and an early study in the Northern Rockies noted that bark-peeled trees occurred in virtually every valley in western Montana and northern Idaho.[16] Only a small proportion of bark-peeled pines are likely alive today because peeled trees were typically large and located on accessible terrain, making them easy targets for timber harvest.

While bark-peeling may not have been as extensive in the Southwest, it was still quite common. Work by University of Arizona ecologist Tom Swetnam suggests that some tribes likely used bark only in times of famine, while other cultural groups may have eaten inner bark every year "because it was valued as a sweet or delicacy."[17]

Bark's importance as a food source continued even after the first wave of Euro-American settlement.[18] In an 1881 article published in the *Nez Perce News* of Lewiston, Idaho Territory, Norman Willey related a story about the emergency use of bark by a small band of Nez Perce Indians being pursued by armed settlers: "But the Indians had left, taking the Indian trail across the divide that separates Long Valley and Indian Valley. . . . A visit to their camp indicated that they are entirely destitute of ammunition. They had peeled bark from a great many trees and had been scraping and apparently living on the soft portions of it, but there was not a bone or feather to be found, although game was plenty thereabouts."[19]

Bark-peeling's historical importance to Native Americans is recognized at several designated cultural sites. The Bitterroot National Forest's Indian Trees Campground just southwest of Sula, Montana, features old-growth pines that were peeled by Salish and Nez Perce Indians 100 to 300 years ago. This bark-peeling site is close to Ross's Hole, the location where Lewis and Clark encountered the Salish on September 4, 1805. However, the first written account of bark-peeling was William Clark's journal entry a week later (September 11, 1805), where he observed this activity along Travelers Rest Creek [now Lolo Creek] about 60 miles to the north: "the Pine trees, pealed as high up as a Man could reach, which we suppose the Natives had done in order to get the inside bark, for to mix with their dried fruit to eat." [20]

Meriwether Lewis also referenced Indian bark-peeling the following spring as the expedition made its way along the Clearwater River in present-day Idaho, with his entry on May, 8, 1806: "During this month the natives also peal this pine and eat the succulent or inner bark." [21]

Another impressive collection of Indian-peeled trees lies in a most surprising place—adjacent to the Great Sand Dunes in south-central Colorado. Here, nearly one hundred ancient pines bear the scars of bark-peeling by the Southern Ute and Jicarilla Apache Indians in centuries past. The Southern Utes lived in the surrounding San Luis Valley and the Jicarilla Apache traveled north from New Mexico to use the area seasonally. Both tribes harvested the inner bark for food while camped at the dunes, but they also used it for medicinal purposes. This small island of living artifacts has been formally recognized by the National Park Service as Indian Grove, making it the only grove of trees on the National Register of Historic Places.[22] Another stand of old bark-peeled pines, called Indian Village Grove, is registered on the list of Oregon Heritage Trees.[23] This grove is located northeast of Enterprise along the route of the Nez Perce National Historic Trail.

Bark wasn't the only component of ponderosa pine that Native Americans used for food. Pine seeds and the light-green, filamentous lichen often found growing on older pines were also occasionally used for food, particularly during times of shortage. Meriwether Lewis mentioned both in his May 8, 1806, journal entry:

> Near this camp I observed many pine trees which appear to have been cut down about that season which they inform us was done in order to collect the seed of the longleafed pine which in those moments of distress also furnishes an article of food; the seed of this speceis of pine is about the size and much the shape of the seed of the large sunflower; they are nutricious and not unpleasent when roasted or boiled . . . we are informed that the natives in this quarter were much distressed for food in the course of the last winter; they were compelled to collect the moss [actually a lichen] which grows on the pine which they boiled and eat. [24]

The wood of ponderosa pine was readily available for many uses, including supports for wickiups and lean-tos. Wickiups were generally dome-shaped structures, and Indians sometimes used large bark pieces, including those from ponderosa pine, to cover walls and roofs. Pinewood was also used to make digging tools, posts, ladders, canoe paddles, cradleboards, snowshoes, and saddle parts.[25]

Ponderosa needles were used for insulating underground storage pits, making mattresses and pillows, and weaving baskets. They were prepared in decoctions to treat internal hemorrhaging and sometimes used as underarm deodorant.[26] Green, fire-singed ponderosa boughs were used in sweat lodge ceremonies to relieve muscle pain.[27] Ponderosa bark was sometimes used to build fires when Indians traveled in dangerous country, because such fires gave off little smoke, cooled quickly, and prevented the enemy from determining when they were used.[28] Ponderosa pine roots were used in basketry, and root extracts were used to prepare a blue dye.[29]

Of all the uses of ponderosa, pitch was likely employed in the most varied ways. The sticky resin was used to seal seams and cracks, attach rawhide

The Navajo used ponderosa pine supports in the entryways of their hogans.
—*Photo by Edward S. Curtis, 1904, courtesy Charles Deering McCormick Library of Special Collections, Northwestern University Library*

to horses' hooves, and waterproof moccasins and the insides of baskets.[30] It was used for chewing gum and applied as a tonic, poultice, and hair-dressing.[31] Pitch was also used as glue in arrow making, as a fuel for torches, and as a thin layer inside flutes to improve their tone.[32]

And sometimes ponderosas were used in ways that may never be clear. On September 11, 1805, two members of the Lewis and Clark Expedition (Ordway and Whitehouse) observed an old ponderosa pine growing along Lolo Creek whose adornments left them wondering. They wrote, "The narrow bottoms on this Creek is thinly covred with pitch pine passed a large tree on which the natives had a number of Immages drawn on it with paint . . . a white bear Skin hung on the Same tree. we Suppose this to be a place of worship among them."[33]

Groves of large, old pines were favored places for both day-to-day living and more formal undertakings, including negotiations and trade. Washington Territorial Governor Isaac Stevens negotiated with chiefs of the Salish, Kootenai, and Pend d'Oreille tribes for nine days in a grove of old ponderosas at a place the Salish called *Clmé* (trees with cut-off limbs).[34] Their meeting culminated with the signing of the Hellgate Treaty on July 16, 1855. Many of these trees still stand, and their historical significance is recognized as Council Grove State Park, located along the Clark Fork River west of Missoula, Montana.

Idaho pioneer Charles Winkler recalled a conversation he had years earlier with Perry Clark, a member of the Idaho Territorial legislature, who was credited with naming central Idaho's Council Valley in the 1870s: "He was in the valley many times and knew of all the uses made of it by Indians, and from that he gave it its name [Council Valley]. In other words, meeting place of the great leaders. . . . They met under five big yellow pine trees that were nearly in the center of the valley. They would race their horses . . . [and] gamble."[35]

Much has changed in Indian life since Perry Clark's observations. With the confinement of Indians to designated reservations in the late 1800s, many historical uses of ponderosa pine gradually diminished. Since the 1950s, harvesting ponderosa pine from tribal lands and selling it for timber products has been one of the most important contributors to the day-to-day well-being of Native Americans. For large reservations with extensive pine forests, such as the White River Apache in Arizona, Nez Perce in Idaho, Flathead in Montana, Mescalero Apache in New Mexico, Warm Springs in Oregon, and Yakama in Washington, proceeds from the harvest of ponderosa pines contribute substantially to tribal coffers. The tribes also benefit from good-paying jobs in managing timberland and harvesting and milling timber from their ponderosa pine forests.

Signing the Treaty at Council Grove. —*E. S. Paxson,* Governor Stevens' Treaty with the Flatheads (Selish)*, oil on linen, 1914; Missoula County art collection managed by Missoula Art Museum; Photo by Chris Autio, 2000*

A memorable visit by author Carl Fiedler to an old Indian campsite in the remote Powder River country of southeastern Montana illustrates the fundamental changes that have occurred in native life and ponderosa forests over the last 150 years. A ponderosa pine tree was found growing within an old tepee ring, a circle of stones that held the tepee cover tight to the ground. A count of the annual rings on an increment core extracted from the tree shows that it germinated in about 1890. Other abandoned tepee rings and numerous other ponderosas growing in the vicinity are silent reminders that, at this site, history has moved on.

PIONEERS IN THE PINES

It's been said that ponderosa pine outlines the boundaries of the American West.

—Anonymous

ALTHOUGH AN EARLY 1500s FORAY INTO the American Southwest by the Spanish conquistadors came nearly three centuries before Lewis and Clark's reconnaissance of the American Northwest, the motivations were similar. Both were looking to find wealth and expand the territory available for future settlement. These efforts to explore and then colonize the American West—though widely separated in time and space—were closely intertwined with ponderosa pine. This tree provided lumber and landmark, fuel and fence post, oar and inspiration to explorers, settlers, miners, and missionaries as they found their way in this new land. But most of all, ponderosa pine—both tree and forest—provided the resources and setting that helped make the West both alluring and livable.

The first European to encounter ponderosa pine was likely Cabeza de Vaca, who shipwrecked off the Texas coast in 1528 and wandered the Southwest for the next eight years. In a later report, he described pinyon pine and another pine, presumably ponderosa, growing in the mountains.[1] Francisco Vasquez de Coronado's expedition of 1540–1542 traveled through central and northern New Mexico, and the expedition's scribe, Pedro de Castañeda, noted, "There are junipers and pines all over the country." He also described a river they encountered: "The natives crossed it by wooden bridges, made of very long, large, squared pines."[2]

In 1598, explorer Don Juan de Oñate established the first Spanish colony, San Gabriel, in northern New Mexico. While traveling back through western New Mexico, his expedition was beckoned by clumps of ponderosa pine growing at the base of a large cliff, suggesting a water source in an

19

otherwise parched landscape. There was indeed a pool of cool, fresh water by the cliff, and Oñate etched his name and the date (April 16, 1605) on the rock.[3] Over the years, hundreds of other early travelers stopped and carved their names here also. Today, Inscription Rock is part of El Morro National Monument.

The Spaniards' oldest documented use of ponderosa pine for building purposes can still be seen today in New Mexico. The Palace of Governors—a nearly block-long structure in Santa Fe—was built in 1610, making it the oldest continually occupied public building in the United States.[4] It served as the original seat of the Spanish colonial government and is now one of the historical museums surrounding the state capitol in Santa Fe. Although primarily adobe, ponderosa pine timbers form supports and frontispieces. Another equally venerable building, the San Miguel Mission, stands a few blocks away. Construction of the mission also started in 1610 but wasn't completed until 1626.[5] Its hefty ponderosa ceiling rafters (vigas) are complemented by beautifully etched decorative timbers.

Ponderosa pine growing at the base of Inscription Rock, El Morro National Monument, New Mexico.

Ponderosa pine beams (vigas) and etched supports in the ceiling of the San Miguel Mission, Santa Fe, New Mexico. —*Copyrighted photo by Alix Bleus*

The Spanish continued their conquest of the Southwest throughout the 1700s, establishing fortified outposts and building missions. Franciscan priests founded more than forty missions, many of them along the Rio Grande River. However, they used only modest amounts of ponderosa pine in the construction of these adobe buildings, similar to the limited uses of pine in the earlier, adobe structures in Santa Fe.

The next big shift in the Euro-American experience with ponderosa pine would not occur in the Southwest. In 1801, newly elected President Thomas Jefferson began promoting an ambitious effort to explore the land between the Mississippi River and the Pacific Ocean, and Congress funded such an expedition in 1803.[6] The timing of the acquisition of the Louisiana Territory in late 1803 could not have been better, because now the expedition would primarily be exploring American territory—not trespassing on land claimed by France. With little fanfare but high hopes, the Corps of Discovery, later known as the Lewis and Clark Expedition, left St. Louis in the spring of 1804 to explore the upper Missouri River and find a route to the Pacific.

The expedition's first encounter with ponderosa pine came on September 16, 1804, when Clark discovered pine cones mixed in with driftwood near the mouth of the White River in central South Dakota.[7] The "pine burs" that he described identify the cones as ponderosa, which likely drifted downstream from near the river's source far to the west.

The following spring, as the expedition made its way up the Missouri River in present-day eastern Montana, Lewis observed that the nearby plains were devoid of timber, which he recorded as a serious drawback to future settlement. However, on May 11, 1805, Lewis spotted a patch of trees on a hilltop along the river, and Clark made his way up and retrieved some branches. Examining the branches, Lewis saw similarities to the pitch pine of his native Virginia, prompting the men to refer to this unknown tree species (ponderosa pine) as "pitch pine," "long leaf pine," or "long leafed pine" for the remainder of the journey.[8]

Weeks later when the expedition reached the Great Falls of the Missouri, Lewis began assembling a boat kit that he and Thomas Jefferson had painstakingly designed. The explorers sewed animal skins together and stretched them over the prefabricated boat frame. Lewis had planned to use pine pitch to waterproof the seams, but there were no ponderosas to be found. Forced to improvise, he concocted a homemade sealant, or "composition." After launching his cherished boat on July 9, 1805, Lewis observed, "A greater part of the composition had seperated from the skins and left the seams of the boat exposed to the water . . . and to prevent her leaking without pi[t]ch was impossible with us, and to obtain this article was equally impossible. . . . I therefore relinquished all further hope of my favorite boat and ordered her to be sunk."[9]

Ironically, the explorers would encounter ponderosa pine just 30 miles farther up the Missouri from the failed launch site. The expedition continued upriver through the Gates of the Mountains, and then left ponderosa country near present-day Helena to ascend high valleys and the crest of the Rockies before dropping into the Bitterroot Valley with its impressive ponderosas. Although Lewis and Clark provided little detail about the forests they encountered, thirty-eight years later, in 1843, a German adventurer and naturalist named Charles Andreas Geyer followed the same route into the Bitterroot Valley and gave a vivid description: "The bulk of the woods consists of the majestic and valuable *Pinus ponderosa*, attaining an average height of 150 feet, and not seldom a trunk from 4 to 8 feet diameter, beautifully rounded and clothed with reddish brown bark."[10]

Continuing westward, Lewis and Clark crossed the imposing Bitterroot Mountains before descending into the upper Clearwater River drainage of present-day northern Idaho. Here, a second attempt at floating the expedition would succeed, thanks to the Nez Perce Indians. These people knew ponderosa pine and canoe building, and shared their knowledge with the weak and hungry explorers. On September 25, Clark wrote, "I Set out early with the Chief and 2 young men to hunt Some trees Calculated to build

Canoes . . . we deturmined to go to where the best timbr was and there form a Camp."[11]

The party found a stand of large ponderosas and soon went about falling five of the trees for canoes. But felling the big trees was far easier than transforming them into canoes. Fortunately, the Nez Perce had perfected a method using fire, rather than axes, to hollow out tree trunks. Having learned the technique, expedition member Patrick Gass noted, "All the men are now able to work; but the greater number are very weak. To save them from hard labour, we have adopted the Indian method of burning out the canoes."[12] Soon, members of the expedition were floating westward toward the Pacific in their new, fire-hewn ponderosa pine canoes.

Lewis and Clark collected many plant specimens during their two-year exploration, including ponderosa pine. At least 178 of these species turned out to be plants previously unknown to science. Frederick Pursh, the esteemed botanist who reviewed the expedition's plant collections, did not recognize Lewis's specimen of ponderosa pine as a new species, perhaps because Lewis repeatedly referred to it as long leafed pine in his notes.[13]

Euro-American ventures into the West gradually increased following Lewis and Clark's return in 1806, sparked by reports of a potentially lucrative fur trade. For the next several decades, the West was the domain of itinerants—fur trappers and the occasional naturalist and explorer. One of these was David Douglas, a plant collector from England who collected a branch of a ponderosa pine near present-day Spokane, Washington, in 1826, but didn't recognize it as a new species until later. Although trappers and adventurers such as Douglas traveled widely, their numbers were few. Historians estimate that only about 3,000 such men roamed the West during the peak of the fur trade between 1820 and 1840.

However, just as the fur trade was fading, a new wave of adventurers was arriving. Some were looking to expand their horizons, others their religion. The first of these were missionaries Marcus and Narcissa Whitman, who established a Presbyterian outpost in 1836 at the confluence of the Columbia and Snake Rivers in southeastern Washington. In 1841, a black-robe named Father Pierre-Jean De Smet arrived in western Montana's Bitterroot Valley to establish the first Jesuit mission. He was joined by Father Anthony Ravalli in 1845, who immediately saw the potential for using the valley's large ponderosas for lumber to expand the mission community. The resourceful Ravalli then set out to design a sawmill, complete with circular saw. He heated and pounded flat the iron rim of a wagon wheel, cut teeth into it, and then fashioned a means to turn the blade with water from a nearby stream.[14] This Renaissance man—missionary, physician, architect, sculptor, and now millwright—had just developed the first ponderosa pine sawmill in Montana.

In the late 1830s, more and more settlers began moving west, following a trail first laid by fur trappers and later to be known as the Oregon Trail. One famous landmark along the route was the giant Lone Pine, or *l'arbre seul* (lone tree) as the French-Canadian trappers called it, a solitary ponderosa that stood in a barren valley north of present-day Baker City, Oregon.[15] It was a welcome sight seen for miles by the first wave of approaching wagon trains, and a favored place for emigrants to camp. As pioneer Medorem Crawford wrote in his journal in 1842, it was a landmark "respected by every traviler through this almost Treeless Country."[16] Historian John Evans noted that it also represented something larger, as ponderosa pine more than any other tree represented the end of the plains.[17] But with increased travel along the trail, the Lone Pine was destined to fall to satisfy some pioneer's appetite for easy firewood. Explorer John C. Frémont would pass this way on October 15, 1843, and he observed just such a fate for the famous landmark: "On arriving at the river, we found a tall pine stretched on the ground, which had been felled by some inconsiderate emigrant ax. It had been a beacon on the road for many years past."[18]

Although ponderosa pine played some role in the life of nearly all new Euro-American arrivals into the 1840s, the overall use of ponderosa pine was still minimal. Then overnight, one four-letter word would do more to change the people-pine equation than all other preceding events. That word was *gold*.

In the late 1840s, John Sutter and James Marshall had thrown in on a business venture in California Territory. Sutter had the money and needed lumber to build his dream village of New Helvetia (later Sacramento), and Marshall had the knowledge to build a water-powered sawmill to manufacture the lumber. They were drawn to a location along the South Fork of the American River because of the area's extensive ponderosa pine forests. Then on January 24, 1848, while construction was still underway, Marshall glimpsed some shiny flecks in the mill's tailrace. After a couple of tests, the flecks were found to be gold.

The California Gold Rush was on, and suddenly prodigious amounts of wood were needed for sluice boxes, flumes, mine construction, cabins, bunkhouses, commercial buildings, fuel, bridges, and, later, railroad ties and train trestles. Ponderosa pine was the most accessible tree and had preferred qualities for many uses. It was light in weight, easy to work, and resistant to splitting when nailed. The appetite for wood to build the mining and related infrastructure grew so rapidly in the 1850s that "new sawmills sprang up everywhere in every settled corner."[19] And there was wood to cut. Abraham Cunningham, a gold miner who arrived in northern California in 1849, described forests that "stretched for miles in every direction with great trees . . . one hundred feet to the first limb. From the Manton area to the Whitmore country and from Inwood to the base of the High Sierras stretched a forest primeval, the greatest virgin stand of [ponderosa] pine and sugar pine that the world has ever known."[20]

Some members of the Mormon Battalion had stayed behind after the Mexican-American War and were working at Sutter's sawmill when James Marshall found gold. While initially turning their energies to mining gold, they soon left to rejoin the community of Latter Day Saints in Utah. Their initial effort to travel in early May was turned back by deep snows. Three advance scouts tried again in late June, but all three were killed by Indians a few days later. In memoriam, survivors removed a patch of bark from a large ponderosa pine and carved the victim's names and the date of death into the tree.[21]

Ponderosa pine headstone for three members of the Mormon Battalion. The stump is on display at the Marshall Gold Discovery State Historic Park in Coloma, California. —*Photo courtesy Historic American Buildings Survey, Library of Congress Prints and Photographs Division, Washington, DC*

Ponderosa pine played another, often overlooked role in the California Gold Rush. Not far from Log Springs, near present-day Jenkinson Lake, a 200-foot-tall ponderosa blazed with a right-facing arrow directed thousands of would-be miners to the goldfields. This landmark soon became known as the 49er Tree, and it still stands today along the Mormon Emigrant Trail, although vandals have defaced the blazed arrow.[22]

By late 1849, more than 80,000 people had arrived in California Territory to find their fortune. By 1854, the number of arrivals had swelled to 300,000. Gold seekers came from all parts of the globe and all ethnic backgrounds, ultimately leading to one of the most diverse and technologically advanced populations in the world. The Gold Rush also spurred building of the first transcontinental railroad, a monumental project that powered further development of the West.[23]

Construction of the first transcontinental railroad from Sacramento eastward through the Sierra Nevada started in January 1863 and immediately required huge volumes of wood. Although redwood was used for railroad ties laid west of the Sierra, ponderosa pine and sugar pine were used for the great length of track east of the Sierra crest and for most other construction and fuel. In addition to the use of about 2,500 ties per mile of track, timber was needed to build depots, warehouses, section houses, living quarters, telegraph poles, trestles, tunnel linings, and snow sheds.

Construction supervisors soon learned the hard way that protecting large sections of track from heavy winter snows was critical to keep the trains running. After some experimentation, they settled on a snow-shed design that kept snow off the tracks. Constructing the 37 miles of snow sheds needed to allow winter travel over mountainous sections of track required 65 million board feet of lumber.[24] (A board foot equals the volume of wood in a board 12 inches wide, 12 inches long, and 1 inch thick.) Additional large volumes of wood were needed to fuel the railroad locomotives and stationary steam engines used in construction, all of which were wood burning. A September 4, 1865, article in the *Sacramento Daily Union* described the impacts of the railroad's appetite for wood on construction in Colfax, a growing town along the tracks west of the Sierra: "House building progresses slowly on account of the scarcity of lumber, caused by the fact that several of the mills in the neighborhood are busy in preparing lumber for the Railroad Company."[25]

As railroad construction moved east of the Sierra crest, more than a dozen sawmills were running full steam near the Nevada border to keep up with the railroad's insatiable demand for wood. Trestles along the route also consumed a gigantic quantity of ponderosa pine timber. By 1869, the Central Pacific Railroad had constructed more than thirty bridges and trestles across

California and Nevada. The longest combined trestle and bridge (5,086 feet of trestle and 400 feet of bridge) spanned the American River, while the Secret Town trestle was perhaps the most picturesque.

New demands for timber were emerging from California's growing cities and towns. However, much of the easily accessible timber had been cut and now logging had to reach out farther from populated areas and higher into the mountains, often into rugged terrain. At a time with no mechanized transportation, a primary challenge was how to efficiently transport timber over ever-longer distances. A man named Charles Ellsworth hatched an ingenious solution to this problem, the V-flume. This was a real breakthrough, as historian Dottie Smith credits the V-flume for unlocking the treasures of northern California's sugar and yellow pine forests.[26] The concept was to saw logs into rough lumber at mountain sawmills and then float the lumber on water in a wooden flume to finishing mills in the valley below. Historian William Hutchinson observed, "Each mile of flume required some 135,000 board feet of lumber, which shows why a sawmill that built a flume was its own best customer during construction."[27] The desired grade of 27 feet of drop

The Central Pacific Railroad's picturesque Secret Town trestle near Gold Run, California, was constructed circa 1865 using pine timbers.
—Photo from Alfred A. Hart Stereographic Collection, courtesy Bancroft Library, University of California, Berkeley

per mile of flume couldn't always be maintained. At one curve in Ellsworth's flume, the drop was so abrupt that lumber reached a speed of 50 miles per hour before the grade tapered off again.

Like so many original ideas, the 40-mile flume was initially mocked and became known as Ellsworth's Folly. Tragically, Ellsworth never saw the end results of his visionary system, as he fell from the flume while supervising final construction and died from his injuries.[28] If imitation is the sincerest form of flattery, then Ellsworth's Folly was a posthumous success. In 1873, construction started on an even longer V-flume known as Blue Ridge Flume, which was supported by trestles as high as 150 feet above the ground.

The V-flume was cleverly designed to self-clear if a jam started, but if it didn't, a flume tender was on duty to prevent bigger jams from developing. It also took less lumber to construct a V-flume and far less water to operate than the conventional square-box flume. An article in the August 1, 1874, Red Bluff *Weekly Sentinel* boasted that it had "brought down pieces 14 inches square and 40 feet long."[29] Lumber was usually flumed plank by plank, with more valuable grades clamped together in bundles or on top of lower-grade lumber. Transporting lumber by flume was wasteful because the ends of the boards were battered during transport and had to be trimmed after fluming. However, the flume could deliver lumber quickly by any standard. An 1880 book profiling life in Tehama County praised the efficiency that resulted from linking the telegraph system and V-flume: "A builder finds his work at a stand-still for want of a stick of timber; he goes to the office . . . his want is telegraphed to the mill; the log is cut, hauled and sawed, and by night his stick comes booming into the yard! This is lumbering in the Sierras!"[30]

Lumber wasn't the only product sent down the flumes. Pinewood for fuel, wooden bolts for barrel staves, slabs for sash and door manufacture, and even boxes of trout and venison packed in snow were transported by flume.[31] Sick or injured people were also sent down to the valley on a craft known as a flume boat, a wild ride that ensured passengers temporarily forgot the nature of their affliction.

California's ponderosa pine forests provided another important but little-known product during the 1860s. Secession of the southern states in 1861 left the North short of turpentine and resins during the Civil War, products that were essential to its fleet of wooden ships and usually obtained from pines in the southeastern United States. Given the North's predicament, one California historian noted that the magnificent pine forests of the Sierra Nevada "were invaded by men who gouged great holes in the bases of ponderosa pines in order to collect pitch that would be made into turpentine and

Flume tender watching for lumber jams along a V-flume. —*Ink drawing by Will Taylor published in* Sierra Flume and Lumber Company *magazine; image courtesy Meriam Library Special Collections Department, California State University, Chico; drawing on loan from William H. Hutchinson*

resin."[32] Another historian ruefully described the unintended consequences of Civil War turpentine gathering: "All that was left to commemorate the California effort were the rocks embedded in the trees, where they had been placed to plug the holes through which the pitch had been gathered."[33]

Apparently unaware of the collection operations in California's pine forests during the Civil War, forestry researchers initiated an experiment in northern Arizona in 1910 to determine whether ponderosa could provide a good source of turpentine.[34] They sampled both young pines and old-growth pines and different methods of extraction. Whether as a direct result of the Arizona research or his familiarity with the California experience a half century earlier, Theodore Woolsey noted in 1911 that southwestern ponderosa pines produce excellent turpentine and predicted that this by-product could become commercially important.[35] While Woolsey's first assertion is likely true, his prediction about its commercial use did not come to pass.

Just as the start of the gold-mining and lumbering eras roughly coincided with the waning of fur trapping, so too did a new period of military expeditions. This effort, called the Great Reconnaissance, was launched to explore, map, and expedite development of the West. A big first step came in 1838

Turpentine collection experiment on the Coconino National Forest, Arizona, in 1910.
—*Photo courtesy Olberding, Huebner, and Edminster 2007; Fort Valley Experimental Forest historical photographs. Fort Collins, CO: US Forest Service, Rocky Mountain Research Station. http://www.fs.usda.gov/rds/imagedb/*

with creation of a separate branch of the military called the US Army Corps of Topographical Engineers. While the Corps' life span was only twenty-five years, some scholars contend that this small group of explorers did more to win the West than any other group in history.[36] The Corps's expeditions were commissioned to survey and map unexplored regions, establish boundaries, and evaluate possible routes for a transcontinental railroad. Other explorers, such as John Wesley Powell, were privately sponsored or directly funded by Congress. Leaders of these expeditions recorded their tree-related observations and experiences, which often involved ponderosa pine.

In 1849, Lt. James Simpson was traveling west from Santa Fe for a meeting with Navajo leaders to negotiate a peace treaty. On the way, he serendipitously encountered the Anasazi ruins in western New Mexico.[37] Simpson carefully recorded his observations of the pueblos, their pine timbers, and the surrounding landscape. Simpson may have been the first explorer to refer to ponderosa pine as "yellow pine" and described them in utilitarian terms that befitted an engineer: "about 80 feet high and 12 feet in circumference at the trunk."[38]

In 1853, Lt. A. W. Whipple—known as the Pathfinder of the Southwest—was searching for a rail route from Fort Smith, Arkansas, to Los Angeles. While passing along the flanks of Arizona's San Francisco Peaks in December 1853, Whipple wrote, "Now we entered a forest of pines extending over a large tract of country south to north. It is a species of yellow pine. . . . They are tall, straight, and sound; from 1 to 3 feet in diameter, and from 60 to 100 feet in height. They are the same that are used for timber throughout New Mexico."[39]

Whipple and his survey party were probably the first to include ponderosa pine as part of a Christmas light show. On December 24, he wrote, "Christmas eve has been celebrated with considerable éclat. The fireworks were decidedly magnificent. Tall, isolated pines surrounding camp were set on fire. The flames leaped to the tree-tops, and then, dying away, sent up innumerable brilliant sparks." [40]

Lt. Edward Beale led another expedition to the Southwest in 1857, a trip that provided the first field test of twenty-eight camels from the US Camel Corps.[41] While the camels would turn out to be a flop, the pathfinding was successful. Beale's party surveyed a route west from Fort Defiance on the Arizona–New Mexico border, through the ponderosa forests of northern Arizona, to Fort Tejon in California. The finished route, known as the Beale Wagon Road, would become a popular immigrant trail during the 1860s and 1870s and would later be followed by legendary Route 66, the Santa Fe Railway, and I-40. While camped in Arizona, Beale and his men cut the limbs off a tall, straight ponderosa pine so they could hoist the American flag. By popular accounts, this ponderosa pine flagpole became the namesake for the city that later sprang up a few miles to the west: Flagstaff, Arizona.

In the summer of 1869, Civil War veteran Major John Wesley Powell and his men were making the first-ever run down the Colorado River to explore the Grand Canyon and made camp one evening at the mouth of Bright Angel Creek. The crew had lost an oar for their boat and found themselves literally up a river without a paddle. Looking for some sort of alternative, they scrambled their way up Bright Angel Creek and happened upon a ponderosa pine log that had washed down from the North Rim of the canyon. Seizing the opportunity, they dragged the log several miles back to camp and soon were back on the river with a newly crafted ponderosa pine oar.[42]

On July 2, 1874, an up-and-coming Army officer named George Armstrong Custer set out from Fort Lincoln, Dakota Territory, with a thousand-man military expedition to explore the Black Hills.[43] It was likely the most lavishly outfitted and best documented expedition of the century, despite its modest mission of exploring the region and evaluating possible sites for a fort. There was also an element of danger in the trip, since the Black Hills were Lakota Sioux territory based on the 1868 Treaty of Fort Laramie. However, with 110 covered wagons, ten cavalry companies, two infantry companies, and an artillery battery, the expedition could quell any hostilities.[44] The entourage included a company band for evening concerts, bottled wine, and several hundred cattle to ensure fresh meat for fine dining, along with hunting dogs, a botanist, a geologist, a naturalist, a zoologist, doctors, engineers, miners, newspaper correspondents, and even a photographer.[45] The photographer, W. H. Illingworth, took eighty photographs on the trip, many of which included the Black Hills's renowned ponderosa pine forests. Recent photographs taken from the same photopoints show that in nearly all cases, today's ponderosa forests are denser than those photographed in 1874.[46]

Custer's expedition was more than the innocent undertaking it was advertised to be. The alleged primary purpose of the reconnaissance was to evaluate the Black Hills as a location for a military fort. However, the inclusion of a geologist and several miners in the expedition confirms the intent to evaluate the region's mineral potential. Articles two years earlier

The black-and-white photo on the left, taken in 1874, shows Custer's impressive wagon train and an open pine forest in the background. The color photo, taken in 2001, shows much denser forest conditions. The two blackened ponderosa pine stubs visible in the lower right portion of the 1874 photo were still standing in 2001. —*Historical photo by W. H. Illingworth; matching modern photo by Paul Horsted*

in the *Yankton Press* also reflect a lust for timber: "There are other sources of wealth besides the gold. We want the lumber to build our cities and towns, and the . . . abundance of pine there will be an additional inducement to stimulate the enterprise."[47] Also in 1872, a congressman introduced a bill "to purchase the Black Hills pinelands from the Sioux" and if they refused, "the Secretary of War would be authorized to negotiate with them to erect sawmills and send lumber down to the prairie regions by raft."[48]

In early August 1874, Custer sent an expedition progress report to Fort Laramie by horseback. The expedition's most significant finding was not the preferred location for a fort. Instead, in Custer's words, "gold has been found at several places, and it is the belief of those who are giving their attention to this subject that it will be found in paying quantities. I have on my table forty or fifty small particles of pure gold . . . most of it obtained today from one panful of earth."[49]

The news traveled fast. By the time the expedition had returned to Fort Lincoln in late August, the Black Hills gold rush had already begun. Heavy cutting of ponderosa pine soon followed to support the mining effort. This

blatant disregard for the 1868 treaty did not sit well with the Lakota Sioux, and it would contribute to Custer's demise two years later at the Battle of the Little Bighorn.

With the discovery of gold in Idaho and Montana around 1860 and the passage of the Homestead Act in 1862, a thriving steamboat trade developed to bring people and goods from the Midwest up the Missouri River to Fort Benton, Montana Territory. Driftwood and cottonwood trees growing along the bank and on islands were originally used to fuel the boats. However, given the heavy steamboat traffic on the river and a large steamboat's enormous appetite for wood (fifty to seventy-five cords per day), the easy wood was soon depleted. Early entrepreneurs called woodhawks quickly seized the opportunity. They began moving up into the coulees along the river and cutting ponderosa pine. The pine bolts were then brought down to the river, stacked by the cord, and sold to passing steamboats for whatever the market would bear. One such steamboat, the Far West, later served as supply ship, expedition headquarters, and staging area for Custer's troops at the Battle of the Little Bighorn in June 1876.

Also in 1876, but 800 miles to the southwest, a different kind of battle was brewing. Author Dan DeQuille, mentor and friend of Mark Twain, was writing to alert the public about the massive exploitation of pine forests associated with mining the Comstock Lode in western Nevada. In his seminal 1876 book, *History of the Big Bonanza*, DeQuille wrote, "Not less than 80 million feet of timber and lumber are annually consumed on the Comstock lode. . . . At the same time about 250,000 cords of [fuel] wood are consumed. . . . The pine-forests of the Sierra Nevada Mountains are drawn upon for everything in the shape of wood or lumber. . . . For a distance of 50 or 60 miles all the hills of the eastern slope of the Sierras have been to a great extent denuded of trees." In one excerpt rich with hyperbole, DeQuille wrote, "The Comstock lode may truthfully be said to be the tomb of the forests of the Sierras. . . . The immense bodies of timber now being entombed along the Comstock will probably be discovered some thousands of years hence . . . in the shape of huge beds of coal, and the geologists of that day will say that this coal or lignite came from large deposits of driftwood at the bottom of a lake."[50]

DeQuille blamed the advent of the flume for expediting logging in areas that would otherwise have been too distant to access. He also described a waterless flume, called a chute or trough, which was sometime used to transport large pine logs: "In some localities a kind of chute is in use, made by laying down a line of heavy timbers in such shape as to form a sort of trough. . . . When the troughs are steep, the logs rush down at more than railroad speed, leaving behind them a trail of fire and smoke. Such log-ways are generally to be seen about the lakes and are so contrived that the logs leap from them into water of great depth, as otherwise they would be shivered to pieces."[51]

Heavy logging of ponderosa pine in the 1870s was not restricted to the east slope of the Sierra but was a part of life across much of the Southwest. From 1878 to 1880, ponderosa pine forests were heavily logged across broad swathes of northeastern New Mexico to provide ties and timbers for construction of the Atchison, Topeka & Santa Fe Railway. In the 1880s, ponderosa pine stands on the Colorado Plateau from Flagstaff to Williams, Arizona, were logged over to support settlement construction. And by the late 1880s, the Central Arizona Railway was moving large volumes of ponderosa pine by rail to population centers across Arizona.

Despite the fact that both major lumber flumes in California had gone belly-up after just a few years of operation, development of new flumes continued. In 1889, an outfit called the Montrose Placer Mining Company announced a grand scheme to construct a flume high up on the sandstone

Thrill-seeking lumberman riding a ponderosa log down a chute into a lake. —*Ink sketch from Dan DeQuille's 1876* History of the Big Bonanza, *courtesy Nevada Historical Society*

Central Arizona Railway train unloading ponderosa logs at the Arizona Lumber and Timber sawmill in Flagstaff (circa 1890). —*Photo from Cline Library Special Collections, courtesy Northern Arizona University Ecological Restoration Institute*

cliffs along the Dolores and San Miguel Rivers in Colorado. Their intention was to channel 80 million gallons of water a day through the Hanging Flume to hydraulically sluice gold from gravel riverbanks downstream. From 1889 to 1891, wagons worked their way along the river to deliver the over 1 million linear feet of ponderosa pine planks needed to construct the 10-mile flume.[52] Engineers who recently examined the flume's remains are unsure why the dimensions varied throughout its length but speculate that huge knots found in the ponderosa pine planks may have contributed. The $170,000 cost of building the flume exceeded the value of gold recovered, and it went bankrupt in 1893. The Hanging Flume is on Colorado's list of most endangered places and was on the World Monument Fund's 2006 list of 100 Most Endangered Sites.[53]

Another widespread use of wooden flumes in the inland West was to divert flow from mountain streams to irrigate pastures, crops, and hay fields in the valleys below. Millions of board feet of wide boards sawn from the original old-growth ponderosas were used for this purpose, and many of these wooden flumes are still in service.

Meanwhile, in western Montana lumbermen were looking to exploit nature's version of a flume—the river—to carry their ponderosa logs to a downstream sawmill. In 1885–86, the Montana Improvement Company built a sawmill at the mouth of the Big Blackfoot River, adjacent to the recently completed Northern Pacific Railroad.[54] The location was ideal. Huge volumes of virgin ponderosa pine grew upriver, the rapidly growing trade center of Missoula lay a few miles downstream on the Clark Fork River, and the Butte-Anaconda mining complex with its insatiable demand for wood was located 100 miles east by rail at the Clark Fork's headwaters. Timber was abundant, regulations were few, and the plunder was on. For the next three decades, millions of board feet of ponderosa pine were floated downriver to the mill each spring. Big old ponderosas were cut and bucked into logs in winter, and then skidded to the river's edge, where they were stacked, awaiting the spring runoff. The rising river would float the logs and initiate the long downstream journey to the mill. The trip was not a smooth one because the logs had to be shoved, cursed, and dislodged along the way by men called river pigs. Few men worked in more dangerous and uncomfortable conditions than the pigs. Despite getting splashed by icy water and banged up along the way, the river pig's job was to keep the logs moving. When logs piled up like pick-up sticks blocking the river, the pigs risked life and limb crawling out and dynamiting the jams to get the logs flowing again. Similar drives to transport logs were made on many rivers throughout the inland Northwest.

If the river pigs in Montana had their difficulties moving logs on the Blackfoot River, their problems paled in comparison with those faced by early settlers in southern Utah. Mormon pioneers who moved to Zion

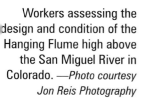

Workers assessing the design and condition of the Hanging Flume high above the San Miguel River in Colorado. —*Photo courtesy Jon Reis Photography*

Log jam on the Blackfoot River, Montana, circa 1910. —*Photo courtesy Archives and Special Collections, Mansfield Library, University of Montana, Missoula*

Canyon in the early 1860s found that cottonwoods, which have poor-quality wood for building, were the only trees available on the canyon floor. A beautiful stand of large yellow pines grew in the area, but on top of a nearby mesa with 2,000-foot vertical cliffs. The only other option was to haul lumber by wagon from the Kaibab forest in northern Arizona, but that alternative involved a ten-day round-trip.

A diary entry made years earlier during a visit by Brigham Young inspired a man named David Flanigan to crack the code of the cliffs. Young had proclaimed that timber would one day come down from the cliffs like a hawk flies. In 1898, Flanigan set out to fulfill what he understood to be a prophecy. The first attempt—dropping a large pine 2,000 feet into sand at the base of the cliff—failed when the tree shattered. He next developed an idea that involved wooden towers and a cable system. Unable to get the necessary financial support, he rigged a poor-man's alternative and ran wire from a windlass at the top of the cliff down to a towerlike structure at the base. In August 1901, after much tinkering with a braking system, a test run of ponderosa slabs was sent down to the canyon floor.[55]

Flanigan then moved a small portable sawmill to the top of the mesa and began sawing ponderosa lumber. A load of pine boards could be moved from top to bottom in only 2.5 minutes, and lumber transported by the cable system was used to construct many of the homes and buildings in the area. Zion National Park was established in 1919 and included the cliff-to-valley cable system. The last noteworthy use of the system was in 1924, when it transported the lumber used in building the original Zion Lodge.[56] Portions of the cable system framework are still visible today on top of the mesa, now called Cable Mountain. Flanigan's cable system was placed on the National Register of Historic Places in 1978. To this day, Zion National Park retains the distinction of being the only national park to have supported an active ponderosa pine sawmill.

Ponderosa pine also played a timely role in another western national park, but this time in its creation. In 1904, a Canadian-born artist named Julian Itter moved to Seattle, and the following summer he spent time painting scenes in the Lake Chelan area amidst the rugged peaks of the North Cascades. This move would prove to be a life-altering experience, with ramifications far beyond any he could have anticipated. In 1906, Itter's paintings were featured in an exhibition at the Butler Hotel in Seattle, and one painting in particular, *Ponderosa Pines in the North Cascades*, received rave reviews.[57] Both the Seattle Chamber of Commerce and the Mazamas, an Oregon mountaineering group, championed Itter's work. Leveraging his popularity, Itter traveled to Washington, DC, and visited President Teddy Roosevelt, to whom he proposed creating a national park that would include Lake Chelan, Glacier Peak, and the land between Cascade Pass and the Skagit River. He later returned and again painted at Lake Chelan. Itter's paintings and dedication to the North Cascades and Lake Chelan area earned him recognition as a significant conservation figure. The area was formally established as a national park in 1968, and Julian Itter would become known as the founding father of North Cascades National Park.[58]

By the 1910s, fledgling management of national forests had begun, and internal combustion engines and electric motors gradually began replacing animals, water, and steam engines in the harvest, transport, and processing of ponderosa pine. The pioneer era in ponderosa forests had ended.

A SPECIAL TREE

Ponderosa pine surpasses all its race . . . an emblem of strength, it appears as enduring as the rocks, above which it raises its noble shafts and stately crowns.

—Charles Sprague Sargent, *The Silva of North America*, 1897

SCOTTISH EXPLORER DAVID DOUGLAS, a botanist for the Royal Horticultural Society of London, collected a branch of a pine tree near present-day Spokane, Washington, in 1826 because he was intrigued by the mistletoe growing on it. At the time, Douglas thought the tree was a red pine (*Pinus resinosa*; a two-needle pine native to the northeastern United States) and labeled it as such in his field notes.[1] It wasn't until 1829 that Douglas recognized the pine as being a new species. Impressed not only by the tree's great size but also its majesty, he named it *Pinus ponderosa*, after the Latin *ponderosus*, meaning heavy, weighty, and significant.

Ponderosa pine has distinctive tufts of 3- to 8-inch-long needles, borne in bundles of two or three, on stout, spreading limbs. Fist-size, purplish cones cling in clusters near the ends of branches. On ripening they dry out and turn brown, their scales expanding to release a rain of winged seeds that drift in the breeze as they fall to the ground. Anyone who picks up the cones instantly learns they are armed with prickles and need to be handled carefully.

Ponderosa pine's platy, cinnamon-colored bark, which flakes off in jigsaw puzzle–shaped pieces, develops slowly. Younger trees, called bull pine or blackjack pine, have rough, dark-brown to almost black, deeply furrowed bark. Their wood is coarse, stringy, and of low value for lumber because it warps when dried. Newcomers often assume these young trees are a different species from the smooth-barked, old ponderosas that produce fine-grained, high-quality boards. After about 100 years of growth, the

Massive ponderosa pine in Kings Canyon National Park, California.

Ponderosa foliage and mature cones.
—*Photo courtesy National Park Service*

Ponderosa pine foliage, cone, and seeds. —*Drawing from USDA*

dark-brown fissures in the bark of bull pines start to separate, and narrow, reddish-brown plates form between them. As more decades pass, the bark plates become broader, yellowish to orangish in color, and smoother, and the quality of the wood improves. These mature ponderosas are known as yellow pines or yellowbellies.

Hardy, old ponderosas with their colorful trunks hoisting great, out-stretched limbs epitomize both majesty and endurance. Yet old ponderosas across the West differ dramatically in stature, from soaring 250-foot-tall trees in sheltered Oregon valleys to stunted 35-foot veterans eking out an exis-tence in the arid, wind-lashed country east of the Rocky Mountains.

Research shows that despite reasonably similar appearances, pondero-sas across the West differ genetically. One outward sign of this divergence is that ponderosas east of the Rockies tend to bear needles in clusters of two, while trees growing west of the Rocky Mountains typically have clusters of three needles. A subtle but more profound difference in needles between these two varieties of ponderosa lies in the stomates, or tiny pores in the needles that control the exchange of carbon dioxide and water vapor critical to photosynthesis and water balance. Trees in the dry environs east of the

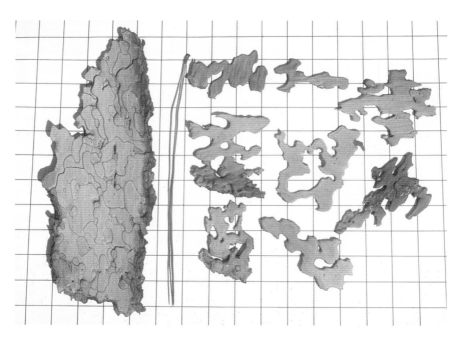

Ponderosa bark flakes look like pieces of a jigsaw puzzle.

Rockies are more efficient at using and conserving scarce water, whereas the three-needle western variety is better adapted for growth. One reason for ponderosa's success across its distribution from Canada to Mexico, the Great Plains to the Pacific, and from near sea level to 10,000 feet lies in its genetically diverse populations, which can be thought of as geographical races.[2] Based on detailed analyses of ponderosa throughout its range, geneticist Robert Callaham recently proposed recognizing five regionally based subspecies as well as transitional forms in some of the areas between them.[3]

Expression of the genetic differences is sometimes startling—just because a race of ponderosa pines can grow in one location does not mean it can grow in another. For example, starting in 1911, foresters collected ponderosa pine seeds from all over the West and planted them in a moist and fertile cultivated field at Priest River Experimental Forest in northern Idaho. Seventy years later, the contrasts were astounding. Pines from various locations in the inland Northwest were tall and straight with healthy foliage. Trees from a South Dakota seed source were dead, although the remaining snags revealed that their early growth had been good. Most trees from seed sources in Colorado and the Southwest grew poorly and died. A similar experiment was also initiated in 1911 near Flagstaff, Arizona. Seeds from eighteen national forests across the West were planted, but after fifty years, seed sources from only nine of the forests survived. Local seed from the Coconino National Forest did best over the course of the study. Overall, the farther the seed source was from the study area, the more poorly the trees did. These results indicate that ponderosa pine's adaptations to local conditions are critical for its long-term success.[4]

Genetic diversity isn't the only reason ponderosa is so widely represented throughout the West; ice ages, humans, and birds played a role, too. During the most recent Pleistocene ice age, which ended about 12,000 years ago, the cold climate displaced forests southward and to low elevations. Afterward they migrated slowly to their modern distributions, but this process was subject to chance. For instance, deserts separate and surround many mountain ranges that provide the only habitats moist enough for trees, and this discontinuity hampers migration. Humans and birds transported seeds across many of these gaps. Native Americans, who used the seeds and inner bark for food, are thought to have transported and planted ponderosa seeds in areas outside their native range, including perhaps Oregon's Willamette Valley and the Fort Lewis prairie near Tacoma, Washington. The jaylike Clark's nutcracker harvests thousands of ponderosa pine seeds and caches many of them in the soil as far as 20 miles away.[5] Stashes of seed that the birds fail to retrieve can ultimately produce new trees.

Newly germinated ponderosa pines from a nutcracker or rodent seed cache.

Ponderosa pine's distribution across the West has a sort of Bermuda Triangle in the center, where ponderosas are missing. Ponderosa is absent or rare in the cold desert country of central and northern Nevada, southeastern Oregon, southern and eastern Idaho, southwestern Montana, western Wyoming, and northwestern Utah. This region is too dry for all forest trees at lower elevations and evidently too cold for ponderosa pine at sufficiently moist, higher elevations, where the frost-free season is short or nonexistent. Mountains in this area only support a forest at high elevations, composed of extremely cold-tolerant trees, such as inland Douglas-fir and lodgepole pine, or they have no forest at all.

Extremely cold winter temperatures don't necessarily exclude ponderosa pine, since it prospers in many places that sometimes plummet to -50 degrees F. Summer frost seems the more likely culprit. Just outside the ponderosa gap region, an area near Chemult in south-central Oregon clearly demonstrates this intolerance to summer frost. Chemult, which sits in a mosaic of flat ground and elevated humps and terraces, experiences frost in every month of the year. Beautiful ponderosas dominate the slightly elevated ground, but they give way entirely to the lodgepole pine in the low-lying places where frost intensifies. Additional evidence of ponderosa's sensitivity to an inadequate growing season is seen throughout high mountains of the West. A hiker ascending the slopes will see that ponderosa pine does not grow as far up as Douglas-fir, lodgepole pine, or other common trees. Also, the uppermost ponderosas typically have frost-damaged tops.

Although ponderosa pine is not found in high-elevation forests, it extends fairly high up on sunny south- and west-facing slopes, sometimes reaching nearly 10,000 feet in the Southwest.

Another way that extreme cold restricts ponderosa pine plays out on the east slope of the Rockies in northern Montana. Ponderosa inhabits the mountains and breaks near the Missouri River in the central part of the state but is not found along the Rocky Mountain Front in northern Montana, and it is absent in Alberta. In contrast, other forest trees found in central Montana do reach north into Alberta. The Rocky Mountain Front is exposed to severe cold waves that sweep down from the Arctic. The town of Browning, east of Glacier National Park, has long held the record for a daily temperature drop, plummeting 100 degrees from 44 degrees F to -56 degrees F on January 23, 1916.

Yet ironically, this region of Montana and southern Alberta is called the chinook belt for the warm winter winds that blast in from the West with dried-out Pacific air masses. These winds, known as snow eaters, warm up as they pass over the mountains and descend to the plains. Chinook winds with

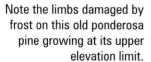

Note the limbs damaged by frost on this old ponderosa pine growing at its upper elevation limit.

temperatures in the 50s (F) can replace frigid Arctic air of -20 to -40 degrees F within hours. Loma, on the upper Missouri River northeast of Great Falls, holds the record for rapid temperature rise—103 degrees F from -54 to +49 on January 15, 1972. These drastic fluctuations can kill foliage of evergreen trees, an injury called red belt. Warm, sunny weather in winter heats the dark foliage while the ground is frozen and roots are unable to replenish leaf moisture lost to evaporation. In the chinook belt, even ponderosa pines that are planted and watered do not survive, nor do the other trees whose natural distributions are confined to the more moderate climates west of the Rocky Mountain divide, including western redcedar, grand fir, and western larch.[6]

In many areas of the inland West, ponderosa pines form a handsome forest in places too droughty to support any other forest tree. Plant physiologists have discovered that ponderosa pine is superior in accessing and conserving moisture. In one historic study, one-year-old ponderosa seedlings attained a height of only 3 inches but developed taproots nearly 24 inches long.[7] Because of these extraordinary taproots, gardeners find it difficult to successfully transplant wild ponderosa saplings.

Roots of two ponderosa pines exposed by erosion along a roadcut in southern Idaho.

This large old ponderosa, growing at the dry edge of its range, towers over pinyon-juniper woodlands north of Flagstaff, Arizona. It illustrates pine's ability to grow to full size for any given location, provided it can overcome the obstacle of initial establishment.

Ponderosa's deep and spreading root systems can efficiently tap moisture far down in the soil. This rooting habit may help explain why ponderosa is able to grow to its maximum size for any given locale, even at the dry edge of its range, a phenomenon Carl Fiedler has observed repeatedly across the West. If a ponderosa can pass the major obstacle of initial establishment, it can generally grow to full size. Sometimes these expansive root systems can be seen when exposed in cut banks along roads and rivers or in deeply eroded cliffs. The dwarfed, often bizarrely shaped ponderosas found growing in rock or on lava flows illustrate the effect of severely constrained rooting.

Ponderosas also have other specialized features to help them survive extreme summer conditions. For example, to prevent water loss during hot weather, they close pores (stomates) in their needles more effectively than other conifers. And when sun beats down on a dark ground surface, ponderosa pine seedlings are able to withstand temperatures up to 162 degrees F.[8] These attributes allow ponderosas to colonize stony and sandy soils that dry out to lethal levels in the upper 2 to 3 feet but remain moist farther down. Central Oregon's Lost Forest, northeast of Christmas Valley, provides

These dwarf pines, growing on exposed rock in Utah, are unable to develop large root systems. The ponderosas in the background, growing in forest soils, measure up to 2 feet in diameter and 75 feet tall.

a dramatic example. Here, a pure stand of ponderosa pine thrives amidst a vast desert where annual precipitation averages less than 10 inches.

One puzzling variation on the dry-site theme warrants scrutiny. Big, spreading black cottonwood trees accompanied by still taller ponderosa pines make up the bottomland forest along some major rivers in the inland Northwest. The deep floodplain soils are saturated with water and often flood during spring. These sites are clearly moist enough for Douglas-fir and other conifers, but many trees cannot tolerate long periods of waterlogged soil, which cuts off the oxygen supply necessary for root metabolism. As summer progresses, the water table drops far below the rooting zone. Eventually, the upper layers of soil dry out enough that they become too droughty for most trees. Thus, these rich floodplain soils are at first too wet and then too dry for conifers—except for the remarkably adaptable ponderosa pine.

Another floodplain forest where ponderosa pine makes a surprising appearance is found near the pine's northern range limits in the Rocky Mountain Trench of southeastern British Columbia. At Canal Flats, a swampy area that forms the source of the Columbia River, lowland spruce (a natural hybrid of Engelmann spruce and white spruce) is prevalent, along with birch and cottonwoods. Spruce is the only floodplain conifer except for a scattering of large ponderosa pines. In contrast, the surrounding upland slopes are dominated by Douglas-fir.

Superior drought tolerance is a crucial advantage for ponderosa since this tree requires ample growing space to thrive and can be crowded out by competitors, notably inland Douglas-fir, grand fir, and white fir. In near-desert climates and in places where coarse soils don't retain moisture—like the deep pumice in central Oregon—ponderosa pines need not compete with other trees. However, approximately half of the roughly 40 million acres that was once dominated by ponderosa pine is moist enough to also support firs.[9] What then was the secret of ponderosa's success in those mixed forests?

Relict ponderosa stumps have been found on millions of acres of moist habitats now choked with thick forests of smaller firs. The stumps contain scars from multiple fires that the trees survived during their long lives and testify to why the open forest changed dramatically to a dense one. Repeated burning historically killed the young firs and kept forests open, allowing pine to prosper. When frequent low-intensity fires were eliminated and the old pines were logged, firs eventually gained control despite initial ponderosa regeneration.

Ponderosa pines require ample growing room to thrive and become big, long-lived trees. Foresters classify ponderosa as shade intolerant, meaning

Ponderosa pine growing in a black cottonwood bottomland.

Ponderosa pine growing in the semiarid canyon along the Salmon River in Idaho.

that it requires nearly full sunlight to develop from a seedling into a full-size tree. The effectiveness of shade in suppressing growth of young ponderosas can often be seen at the edge of openings, where trees growing in full sun are several times taller than those of the same age growing in the shade. Other species that are shade tolerant, like firs, produce abundant saplings that can grow up beneath a canopy of ponderosas and eventually crowd them out. Shade-tolerant trees are not really shade loving, because they too grow better if given more light and space. However, under shady conditions their foliage can photosynthesize and their roots can extract water and nutrients more effectively than ponderosa pines. Even when young ponderosa pines grow in thickets without competition from other species, they often remain small and spindly, vulnerable to disease, wildfire, or collapse under a load of heavy wet snow—a condition known as snow break. Ponderosa pine forests simply cannot sustain a dense growth of trees.

Ponderosa pine forests of the inland West receive only about one-fourth to one-half as much precipitation as forests in the Midwest, South, East, and

West Coast. In addition, the inland West's atmosphere, or relative humidity, is much drier and subjects trees to drought stress for several weeks or months every year. Many other comparably dry regions of the world support only scrubby woodland or brushland, not tall well-formed trees. Historically, fires periodically thinned the ponderosa forest, preventing a thick growth of weak trees that would ultimately be destroyed by severe fire or epidemics of insects or disease.

Although especially resistant to low-intensity fires, ponderosas are vulnerable to a variety of damage agents, most significantly bark beetles and dwarf mistletoe. Bark beetles are always present at low levels in pine forests, typically killing an occasional individual or a clump of trees. However, during periods when trees are stressed by prolonged drought, beetle populations can build up as the trees' natural defenses decline. Recent beetle epidemics in ponderosa forests are likely due to a combination of lingering drought and vast acreages of overcrowded, stagnating trees. Healthy pines smother invading beetles in a sticky pitch, but low-vigor trees cannot repel them. Attacks by overwhelming numbers of beetles can bring on massive mortality that turns trees brown across whole mountainsides. See more about historical and recent beetle epidemics in chapter 10.

Dramatic height differences in young ponderosas of the same age growing in nearly full shade (lower left) to full sunlight (upper right) in central Montana.

Ponderosa pine infested with mountain pine bark beetles.

Dwarf mistletoe, a small parasitic plant similar to the Christmas mistletoe except that it has tiny leaves, also threatens ponderosa pine. The mistletoe sinks modified roots into branches of the host tree and robs it of nutrients and water. Dwarf mistletoe reproduces by explosively launching sticky seeds into the air, sometimes as far as 20 to 30 feet. Seeds attach to anything they contact as they fly out and fall toward the ground, and if they land on an adjacent tree or another limb of the same tree, the infection spreads. Dense pine stands with trees of many different sizes are especially vulnerable to mistletoe, because the parasite can spread more easily. However, surface fires can scorch and kill the shoots of dwarf mistletoe located in the lower tree canopy. Fires also tend to kill heavily infected trees, as their pitchy clumps of swollen branchlets, called witch's brooms, are highly combustible.

Healthy, open forests allow some ponderosas to attain ages of 500 years or more. Colorado naturalist Enos Mills photographed and described a giant ponderosa near Mesa Verde that was felled by lumbermen in 1903. He counted 1,047 annual growth-rings on its massive stump.[10] Similarly, a 5-foot-diameter ponderosa logged in 1926 north of Missoula, Montana, had lived 1,100 years.[11] Some gigantic ponderosa pines were spared from logging and have survived bark beetles, mistletoe, lightning strikes, violent

Dwarf mistletoe has infected this sapling ponderosa pine in New Mexico.

windstorms, and other hazards. Most of these monarchs show their age. The oldest trees have huge, gnarled limbs and thin, ropy branchlets. The crown is broadly rounded, flat, or spike-topped, reflecting centuries of exposure to battering winds, snow, ice, and lightning strikes. Some ancient ponderosas display telltale vertical or spiral scars gouged in the trunks by a bolt of lightning, which sometimes—though not always—kills the tree. Record-size ponderosas, 7 feet or more in diameter and about 200 feet tall, are found in Washington, Oregon, California, and Idaho.[12] The other western states have slightly smaller but still magnificent, record-size ponderosas.

Hawks, eagles, and other raptors build stick-pile nests among the stout limbs or on the jagged, broken tops of giant ponderosas. The ample evergreen canopies of old pines also provide sheltered roosts for blue grouse and families of great horned owls. Flammulated owls in particular favor open ponderosa pine forests, and black bears sometimes hibernate in fire-carved hollows at the base of big pines.

Squirrels fell cones from the ponderosa's bountiful crops and cart the heavy cones off one by one to caches in burrows or rotten logs. However, Abert's (or tassel-eared) squirrels, which live exclusively in ponderosa pine forests, do not cache cones but instead feed on various parts of the tree throughout the year, including the needles, seeds, buds, and even inner bark.

Giant ponderosa pine in Arlee, Montana, 1933. —*Photo courtesy US Forest Service*

Red crossbills cling to cones dangling high in the canopy and pry out seeds from between the scales with their scissor-like beaks. On any given acre of ponderosa pine forest, millions of seeds fall from the cones as they ripen, dry out, and open. Open-grown ponderosas often bear abundant cone crops, while trees in dense stands and those growing in certain regions produce few if any cones most years.

Buoyed by a membranous wing, the seeds can be carried 100 yards in a gust of wind. Once on the ground, many seeds are eaten or dispersed by flocks of Clark's nutcrackers, jays, and other birds that swoop down to the forest floor for half a minute and then suddenly, on some obscure cue, fly back up into a tree. Wild turkeys sometimes spend weeks in autumn raking the fresh pine needle litter aside with their huge feet to find newly deposited ponderosa seeds on the forest floor. Ponderosa pine seeds are larger than seeds of Douglas-fir and most other tree species that live in the same

Bears love to exercise their claws on the smooth trunk of old ponderosas, leaving their signature claw marks as vertical scratches in the bark.

area, and thus are eagerly sought out by a variety of birds, as well as squirrels, chipmunks, deer mice, and other rodents—critters that subsequently become food for foxes, coyotes, hawks, owls, and other predators. Ponderosa pine forests also provide deer and elk with a warmer, less-snowy winter habitat than other kinds of forests. Open-grown ponderosa pines commonly feature vigorous bunchgrass undergrowth, which is more accessible and nutritious than vegetation beneath other forests.

When fire injures a ponderosa pine, the tree promptly seals off the damaged area with a flow of pitch. Because of this natural preservative, when the tree eventually dies, it can be extremely durable. The pitchy, rot-resistant base of the snags allows them to remain standing for up to a century. Old-growth ponderosa snags furnish prime habitat for many species of cavity-nesting birds and mammals—including woodpeckers, songbirds, tree squirrels, flying squirrels, and raccoons. Woodpeckers prefer old pines

A great gray owl surveys the landscape from the branch of a large ponderosa. —*Photo by Lance Schelvan*

A northern flicker makes its home in the trunk of a ponderosa pine. Note all the holes drilled into the trunk by birds feeding on bark beetle larvae. —*Photo by Bonnie Arno*

for excavating their cavity nests because trunk rot tends to be localized and surrounded by sound wood. Abandoned woodpecker nests are adopted by mountain bluebirds and a host of other avians, as well as squirrels.

Unfortunately, ancient snags are a dwindling feature of ponderosa pine forests. Big, old, pitch-filled pines are seldom produced under modern conditions of dense stands and short-lived trees growing in the absence of frequent, low-intensity fires. Many people would like to see the modern ponderosa forest resemble the historical one—to benefit wildlife, reduce the threat of severe wildfire, and once more live among these giants—but to restore sustainable pine forests we must first learn how they developed and were perpetuated.

FORESTS BORN OF FIRE

Wherever it grows, ponderosa pine is born of fire.

—Ronald M. Lanner, *Trees of the Great Basin*, 1983

TO UNDERSTAND THE HISTORICAL PONDEROSA forest requires study, but how can modern scientists do that when the forests have mostly been cut down or overgrown with firs? It turns out that many modern forests preserve a record of the historical forests because ponderosa has unusually rot-resistant stumps and snags. Fires stimulate the concentration of pitch in the base of the trunk and basal roots of a tree. People conducting prescribed burns in ponderosa forests have witnessed a stream of pitch spurting out of a crack or scar at the base of a tree in response to the fire's heat.

Pitch-saturated wood at the base of the trunks produces durable standing snags and rock-hard pitch stumps. Scientists can identify which stumps were once ponderosas because their century-old stumps still contain light-colored, fresh-smelling pitchy wood, and chunks of the ponderosa bark, with its characteristic jigsaw-puzzle flakes, can be found near the ground line. Old stumps from other species of large trees also can be identified from distinctive bark and wood features even when rotten. Thus, the original forests that were logged in the late 1800s can be largely reconstructed based on remnant stumps. Most ponderosa pines that have grown up in modern times, without fires, contain far less pitch wood. When cut, their stumps rot away within a decade or so.

The old stumps, as well as old living trees, often preserve fire scars that are records of each passing fire. Fire scars appear as upside-down Us or Vs and extend to the ground, but they can be narrow or broad, short or tall. Fire scars on ponderosa pines that are 200 or more years old often show up as heavily charred wood, called a cat face, as a result of multiple fires. The first fire kills the cambium (growing tissue), usually on the uphill or leeward side

Burning pitch streams from a scar at the base of a living ponderosa pine.

This century-old ponderosa pine pitch stump has multiple fire scars.

A remnant stump and log of a historical ponderosa pine forest, now replaced by diseased firs.

of the tree, where fallen needles and bark pieces have accumulated. Subsequent fires then burn into the wound's exposed deadwood. Very old trees that were exposed to frequent surface fires in past centuries show countable scars from many different fires.

Reconstructions of fire history from a great many ponderosa forests west of the Rocky Mountain crest, or continental divide, reveal a pattern of frequent fires extending back several centuries and ending in the early 1900s, when wildfires began to be suppressed. Although many trees were not scarred by the low-intensity fires, some survivors exhibit a datable fire scar where the growing tissue was killed by the heat of the fire on one side of the trunk. Thereafter, annual growth rings were no longer produced on this dead segment of the tree's base. However, the adjacent undamaged area of the trunk continued to grow, and its rings slowly expanded over the edge of the dead area. This pattern becomes obvious on a cross section sawn from the fire-scarred trunk. If additional fires burned the scarred tree, they likely charred and enlarged the earlier wound, resulting in distinctive patterns that allow scientists to date the year of each individual fire scar.

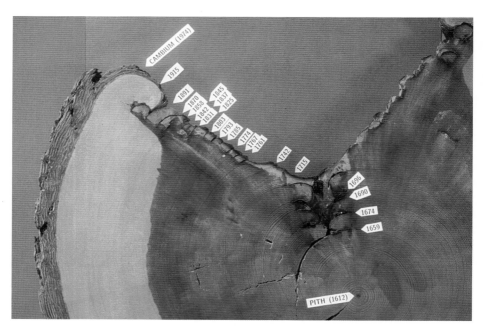

This cross section of a ponderosa pine tree, which began growing in 1612, shows twenty-one fire scars between 1659 and 1915.

A low-intensity surface fire burns through a ponderosa pine forest.

Old-growth trees and ancient pitch-filled stumps in many ponderosa forests throughout the western United States may contain scars from twenty or more fires during a span of two to four centuries prior to 1900. Additional fire dates recorded on neighboring trees indicate that even the most repeatedly scarred trees didn't register all the fires, probably because the pine-needle litter around a given tree trunk was too sparse. Intervals between fires in these forests mostly ranged from an average of only about two years in parts of northern Arizona to twenty-five or thirty years at higher elevations and moist sites, with many areas averaging between seven and fifteen years. This pattern of frequent fires was instrumental in producing and maintaining parklike ponderosa forests with big trees and open, grassy understories. Fires thinned out saplings and shrubs and killed some of the overstory trees, particularly those with a scar exposing heart rot.

In some areas east of the continental divide, from central Montana to the Black Hills of South Dakota and Colorado's Front Range, ponderosa forests burned less often, and some of these were stand-replacing burns that killed clumps or patches of trees. Such areas usually have few old-growth ponderosas or their stumps and are populated by younger pines that grew

This ponderosa pine forest is being replaced by fir because surface fire has been eliminated. Note the fire scar at the base of the large ponderosa.

after a relatively severe fire. These dense young stands occasionally go up in flames, only to be replaced by another "dog-hair" stand, much like the crowded lodgepole pine forests in Yellowstone Park and other high-elevation areas. Also, historical photos reveal that some of the young ponderosa forests east of the Rockies were once grassland with just a few scattered trees or small patchy groves, often confined to rocky areas. Before heavy livestock grazing was initiated in the late 1800s, heavy grass supported frequently recurring fires that killed invading tree seedlings. After a century of fire suppression and grazing, however, young ponderosas are gradually colonizing the edge of dry meadows and grasslands across the West.

In forests that experienced frequent fires prior to 1900, scientists have wondered how far into the distant past this pattern shaped ponderosa forests. Fire-scar records from living pines and old stumps extend back to the 1600s in many areas and to about 1500 in a few places, but this barely predates the influence of European explorers in North America. Studies of charcoal layers in sediment cores taken from ponds and bogs suggest that patterns of frequent burning prevailed for a millennium or longer, but these sources do not provide a definitive record of low-intensity fires.

Conveniently, giant sequoia trees 2,000 to 3,000 years old are part of the ponderosa pine–mixed conifer forest in California's Sierra Nevada. Stumps of giant sequoias logged more than a century ago are rot-resistant and record patterns of frequent fires similar to those seen on the ponderosas. Tom Swetnam and colleagues from the University of Arizona's Laboratory of Tree-Ring Research were able to date fire scars on the sequoia stumps in several different groves, and they discovered that the sequence of frequent surface fires extends back more than 2,000 years.[1]

How was it possible to have so much fire on the landscape? Nineteenth-century photos of the inland West and reliable written accounts, such as those of Lewis and Clark, confirm that the dry regions, including ponderosa pine forests, were vast, unbroken landscapes covered with grass. During an 1853 to 1855 exploration of forests on the east slope of the Washington Cascades near Mt. Adams, J. G. Cooper wrote, "One pine almost exclusively prevails, (*P. ponderosa*, called 'Yellow pine,') growing usually over 100 feet high, with a straight clear trunk 3 to 5 feet thick. . . . There is so little

Ponderosa pine saplings colonize a meadow in western Montana. —*Photo by Lance Schelvan*

underbrush in these forests that a wagon may be drawn through them with-out difficulty. . . . The level terraces, covered everywhere with good grass and shaded by fine symmetrical trees of great size, through whose open, light foliage the sun's rays penetrate with agreeable mildness, give to these forests the appearance of an immense ornamental park."[2]

Ponderosa pine produces large quantities of kindling—the fluffy accu-mulation of pine needles—that promote ignition and allow flames to spread readily through its forests. The grasses that grow in the open forest dry out each summer and also help spread fires. The climate is semiarid, with dry soil and atmosphere, so the ground litter is dry, too. Furthermore, all parts of the ponderosa pine tree from roots to trunk, bark, branches, needles, and cones are permeated with flammable resins and pitch.

Thousands of lightning strikes ignite fires each year across the West, and prior to settlement by European-Americans, nobody put them out. Moreover, flames could travel unimpeded across the fuel-carpeted land because there were no highways, irrigation, dry-land farming, livestock grazing, and all manner of developments to impede the fire's progress. Rivers aren't much of a barrier to fires because wind can loft burning embers across them. Fire

Frequent surface fires maintained the openness of this ponderosa pine forest in western Montana, photographed in 1897. —*Photo by John Leiberg, US Geological Survey*

scientists estimate that in an average year prior to 1900, fires burned 20 to 25 million acres of forest, grass, and sagebrush country in the Mountain and Pacific states, an area five to ten times larger than wildfires burn today, and twenty times larger than what fires burned from about 1930 to 1980 when suppression was most effective.[3]

It is often said that on a clear day in the West you can see forever, but that's not actually true when fires are burning. In the nineteenth century, Major John Wesley Powell, head of the US Geological Survey, portrayed the West as a region of smoke and fire. Traversing from the Dakotas to Oregon and Washington by train in 1889 he remarked, "Among the valleys, with mountains on every side, during all that trip a mountain was never seen. This was because the fires in the mountains created such a smoke that the whole country was enveloped by it and hidden from view. That has been the experience [of my surveying crews] for twenty-odd years, year by year, in this region."[4]

Of course some of those fires can be attributed to the influx of settlers and prospectors, and sparks from the newly established railroads. However, these new human ignition sources were largely supplanting fires set by Native Americans. For example, in 1805 and 1806, before trappers, traders, or pioneers had significantly penetrated the Northwest, Lewis and Clark reported ten fires in what is now Montana and Idaho and attributed seven of them to Native Americans. In the last few decades, investigations of tribal traditions and research by historians, ethnobotanists, anthropologists, archaeologists, and ecologists have revealed a wealth of evidence implying that fires ignited by Native peoples have influenced landscape vegetation for hundreds and probably thousands of years.[5]

Indians used fire regularly for heating and cooking but also for myriad other purposes—burning to rejuvenate forage plants and attract deer, driving game for hunting, clearing campgrounds to prevent ambush, driving enemies out of heavy cover, stimulating food plants and willow sprouts used for basketry, and clearing travel routes. Lewis and Clark and other early journalists noted that Indians often lit fires to signal over long distances, and these spread freely, burning large areas. Native people were often casual in using fire and not concerned about confining it to a certain area. For instance, on August 31, 1805, in the vicinity of what is now Salmon, Idaho, William Clark wrote, "Praries or open Valies on fire in Several places—The Countery is Set on fire for the purpose of Collecting the different bands [of Indians]."[6]

Up until about the 1970s, a prevailing viewpoint held that Native Americans, including their fires, had minimal impact on the environment in the West, primarily because we thought there were so few people living in such a large area. However, the Native American populations encountered by

Lewis and Clark and other early explorers were only a fraction of earlier populations, having already suffered catastrophic die-offs from smallpox epidemics and other plagues brought to North America by Europeans. Books like *1491: New Revelations of the Americas before Columbus* by Charles Mann, *Tending the Wild* by M. Kat Anderson, and *One Vast Winter Count* by Colin Calloway reveal that Native peoples had significantly shaped the wilderness that the first Europeans encountered when they entered the West.

Some people think that the effects of Native Americans on the land shouldn't be considered part of a natural ecosystem. However, people inhabited western North America at least going back to the last ice age— likely for more than 12,000 years. They probably moved into some regions recently devoid of ice at about the same time the first ponderosas germinated from seed there. Increasingly, anthropologists, other natural scientists, and land management agencies, including the National Park Service, consider the effects of native peoples to be part of the primeval conditions. Moreover, in the case of what was natural in ponderosa pine forests, the role of Native Americans may be a moot point because lightning also produced fires aplenty.

What a paradox then, that fire—a tree-killing agent—was instrumental in creating splendid forests covering millions of acres in the Mountain and Pacific States. But how is fire beneficial to ponderosa pine forests? Most of the world's big-tree forests occur in humid climates, but ponderosa pine doesn't fit this mold. It forms stately, enduring trees in areas of deficient rainfall, poor soils, and hot, desiccating summers. Rapid taproot growth and tough, drought-resistant needles allow its seedlings to survive where other tree species fail. But as is common in other forests, often too many seedlings survive and compete with each other for moisture, nutrients, and sunlight. Some mechanism is needed to thin out the small trees. Similarly, something besides fungal decay is needed to recycle the tons of needles, cones, and dead branches and stems that accumulate on the forest floor every year. Decay alone isn't up to the task. Where ponderosa grows, moisture adequate for fungal activity primarily occurs at a time of year when temperatures are too low, and when it is warm enough, it is commonly too dry. Ants help break down deadwood, but not the fallen needles, cones, and bark flakes.

Frequent fires burning across the ground surface provide the ideal thinning and litter-recycling agent in pine forests. The largest, fastest-growing ponderosa pine saplings are the most likely to survive a surface fire, due to thicker bark and foliage that is higher off the ground. In moister, pine–mixed conifer forests where Douglas-fir, white fir, or grand fir can crowd out ponderosa pines, thin-barked fir saplings with foliage near the ground are less likely to survive a fire. Even the larger fir trees mixed with ponderosas can

This dense growth of young pines is likely to stagnate.

be thinned out by fires that damage their shallow roots or thin bark. By consuming the litter layer, fire produces a flush of available nitrogen and other nutrients that contribute to plant and tree vigor. Fire also affects soil chemistry in ways that apparently help control root rots that otherwise can spread through the forest, killing many trees. In addition, frequent fires reduced the numbers of overcrowded, slow-growing pine saplings vulnerable to dwarf mistletoe and diseases like stem rust.

Without fires, saplings often proliferate, and plant physiologists have discovered that an understory of young trees can collectively outcompete the older overstory trees. This unexpected effect is called asymmetrical competition and is comparable to asymmetrical warfare, where small groups of guerillas or terrorists have advantages over a large conventional army.[7] Root systems of understory firs or young pines intercept much of the limited moisture and nutrients, severely stressing all trees. This leaves the large trees vulnerable to attack by bark beetles and other insects and diseases, because they no longer are vigorous enough to produce the defensive chemicals that allow them to drown, or pitch out, invading beetles or render foliage unpalatable to needle-eating insects.

So, we can see how frequent fires maintained the giant pines and grassy glades of historical times, but less apparent is how the old trees died and were replaced by new ones. Determining the ages of all the trees in a given area has shown that old-growth ponderosa pine forests are composed of groups of trees of different ages, distributed in an intricate mosaic. In some cases an unusually severe fire, perhaps following mortality from a bark beetle attack or windstorm, killed most of the trees, and a new stand developed with a few surviving veteran trees imbedded within it.

Experience studying these forests over many years reveals that individuals and groups of large trees die periodically. Even more than among people, life spans of the individual ponderosa pines vary tremendously. When a big tree or group of them dies, the opening provides an opportunity for new seedlings to germinate and grow rapidly, due to reduced competition for moisture, nutrients, and sunlight. Successful regeneration is often episodic, occurring infrequently and depending on a series of conditions such as a good seed crop, favorable spring moisture, and perhaps soil laid bare by a fire.

This large ponderosa in southwestern Colorado survived a lightning strike. Note the missing strip of bark that spirals around the trunk.

Because saplings in the open grow faster than those situated beneath live trees, they have a better chance of surviving the next fire. Once the big dead tree sheds its needles and small twigs and the next fire consumes them, the opening would likely accumulate less litter for a future fire, again favoring survival of saplings. When dead trees eventually fall, as in a windstorm, and later burn, a good microsite is created for new seedlings to establish, sheltered by the remains of the downed trunk. Mature pines growing in a straight line in the forest are evidence of this phenomenon.

Ponderosa's multifaceted relationship with and dependency on fire produced the open parklike groves of the distant past. But then the industrious and orderly citizens of the early twentieth century decided it was in our best interest—and the forest's best interest—to put those fires out.

These five old ponderosas grow in a line (one leaning slightly left) at the Desert National Wildlife Refuge in Nevada. They likely germinated where a log burned years ago, or from seeds cached along a log by birds or rodents.

CRUSADERS AGAINST FIRE

No human force can superimpose a theory over the facts of nature, nor, on any appreciable scale can nature be made to fit a mold of human design.

—K. Ross Toole, *Montana: An Uncommon Land*, 1959

STARTING IN 1871, A SUCCESSION OF TERRIFYING forest fires swept through conifer forests of the northern Great Lakes States, wiping out newly established backwoods settlements, sometimes killing hundreds of people at a crack. The horrific fires followed close on the heels of the first railroad access, which enabled a massive wave of logging and immigration of thousands of homesteaders and would-be farmers. The new settlers lit fires to clear away the thick layer of branches and treetops left by loggers, but high winds spread fire far and wide.

Destruction of the North Woods by logging and fire helped inspire the national conservation movement that induced both Democratic and Republican presidents to designate vast forest reserves in the western United States between 1891 and 1909. This unprecedented action was intended to protect forests and watersheds from plunder and destructive fires, most of which were attributed to human activities. The western forests, however, were inherently more fire prone than the North Woods.

In 1889, while six states from Washington to the Dakotas were on the brink of gaining statehood, wildfires raged across this vast region. On August 15 the *New York Times* reported from Portland, Oregon, that "all the northwestern country seems to be burning up in forest fires. The smoke has been so dense in Portland for the last two or three weeks that for a time it was impossible to see far up the street."[1] The newspaper also reported that 1,000 miles to the east along the new Northern Pacific Railroad, fires were burning through the ponderosa pine woodlands around Miles City in Montana

Territory. Flames charred millions of acres of mountain forests, but heavily grazed grasslands surrounding several towns, including Deer Lodge and Helena, Montana, and Boise, Idaho, apparently helped save them by serving as a fuel break.

September 1902 brought another wildfire emergency to the Northwest. In Oregon and Washington, fires set in logging slash and for land clearing, along with campfires left by hunters and berry pickers, blew up when fanned by strong easterly winds. In all, close to 700,000 acres burned, and at least thirty-five people died. Dozens of loggers survived by diving into the waters of Trout Lake, in the ponderosa pine forest north of the Columbia River Gorge, reportedly along with deer, bears, and other wildlife.[2]

From 1902 onward, the agencies in charge of the federal forest reserves dramatically ramped up efforts to control fires. The US Forest Service was established in 1905, the same year Urling Coe arrived in Bend, Oregon. Coe's memoir, *Frontier Doctor*, provides a haunting account of how the newly minted US Forest Service banned the use of fire over the objections of local wisdom:

> When I came to eastern Oregon in 1905, all of the beautiful [ponderosa] pine timber was an open parklike forest. . . . Each summer there were many forest fires, the vast majority of which were caused by lightning. As there was no underbrush, these fires consumed nothing but the dead pine needles, cones, and twigs that had been blown to the ground by the winds. The little blaze, only a few inches high, crept slowly over the ground and cleaned the floor of the forest of all debris. . . . It was these annual fires which had existed for centuries that had produced the beautiful open forests. . . .
>
> No one tried to put these annual fires out, as they were known to be a benefit to the timber. When the big lumber companies began to buy the timber, their representatives in the field saw to it that their holdings were burned over every year. If the lightning did not start enough fires, the timber men started more of them.
>
> [In a few years, government foresters arrived.]
>
> These new rangers were fine young fellows, mostly college men, who had acquired their knowledge of forestry from books and knew nothing about local conditions. They all had the same conviction and that was that fire should be kept out of the timber at all times and at all costs. . . . The experienced timber men on the ground tried to convince them that fire was necessary and beneficial to the Oregon pine, but it was useless.[3]

Ponderosa pines growing in downtown Bend, Oregon, in 1911.
—*Photo courtesy Des Chutes Historical Museum*

Forest rangers were instructed to be watchful for fires and to extinguish them promptly. However, training, techniques, and tools to accomplish fire control were sorely lacking. Clearly not understanding the western landscape and the job facing western forest rangers, the General Land Office stated in 1902 that rangers would be held personally responsible for any fires that were allowed to escape without adequate explanation. "It is not understood why forest fires should get away from rangers, or rather why they do not find them and extinguish them more promptly. It seems reasonable that a ranger provided with a saddle horse and constantly on the move, as is his duty, should discover a fire before it gains much headway."[4] This was despite each ranger being responsible for well over 100,000 acres of rugged country, much of it without trails.

The US Forest Service began as a tiny agency, but its ambitious leader, Gifford Pinchot, was a close personal friend of President Theodore Roosevelt. In 1898, Gifford Pinchot had warned that "like the question of slavery, the question of forest fires may be shelved for some time, at enormous cost in the end, but sooner or later it must be met." Under Pinchot's leadership, the Forest Service claimed fire control as its compelling mission and in 1908 persuaded Congress to set up a unique system, like an open checkbook, to fund whatever fire suppression efforts were needed. With this financial support, the Forest Service promoted and expanded its fire suppression program. However, the agency's assertion that it could control wildfires was soon undermined by the infamous Big Blowup of 1910 in northern Idaho and northwestern Montana.[5] Severe drought, record-setting high winds, and extreme fire behavior produced the 1910 holocaust. In an epic struggle, the first government-mobilized firefighting forces were pitted against the implacable rage of nature, and the story continues to attract chroniclers to this day.[6]

The winter of 1909–1910 brought an ample snowpack to the Northern Rocky Mountains, but unusually warm, dry weather through the spring and summer eventually produced an alarming, region-wide drought. By mid-summer, hundreds of fires were burning in the mountain forests of northern Idaho and western Montana, the majority caused by man. Since there were no effective measures to extinguish so many fires in remote, rugged country, they continued to grow and coalesce. The drought drug on, and then on August 20 a massive dry cold front invaded from the northwest. Relentless, gale-force winds soon produced a wall of flames shooting high above the treetops. Crown fires merged into a gigantic inferno. Its column of extreme heat and inky smoke was blown horizontally, creating a gargantuan blowtorch. This firestorm spawned hurricane-force winds that flattened great swaths of forest even before the flaming front reached them.

The 1910 fire in northern Idaho, known as the Big Blowup, killed most trees in its path. —*Photo courtesy US Forest Service*

Hundreds of firefighters with only hand tools were chased or overrun by the flames, and heroic rescue trains evacuated thousands of people from the small mountain towns in the firestorm's path. Books by Stephen Pyne and Timothy Egan give blow-by-blow accounts of the many desperate struggles to survive, including how heroic ranger Ed Pulaski saved his terrified crew by ordering them into a mine shaft and keeping them there at gunpoint.[7]

Forty years later, northern Idaho resident Betty Goodwin Spencer described the firestorm's onslaught: "The heat of the fire and the masses of flaming gas created whirlwinds that mowed down mile-wide swaths of pine [mostly lodgepole and western white pines] and fir and cedar in advance of the flames. And behind all this, advancing ominously and steadily, destroying everything in its path—the ground fire. . . . Fire brands the size of a man's arm were blasted down in the streets of towns fifty miles from the nearest fire line. The sun was completely obscured in Billings, five hundred miles away from the main path of the fire."[8]

The monstrous conflagration engulfed 3 million acres in a two-day rampage that killed eighty-five people and destroyed small settlements in the heavily forested mountains of northern Idaho and northwestern Montana.[9] This catastrophe cemented and intensified the Forest Service's determination to control the fire fiend.

The Big Blowup's relationship to forest ecology and especially to the West's most fire-prone forest type—ponderosa pine—is seldom clarified and still clouds perceptions of fire's natural role in western forests. The forest consumed in the Big Blowup was primarily fir, cedar, lodgepole pine, and western white pine, growing in a moist mountain region with a history of high-intensity fires occurring at long intervals—100 years or more. Stephen Arno has observed scars on surviving trees that indicate where the 1910 fires spread eastward into drier forests dominated by ponderosa pine, they often burned at lower intensity, resembling pre-1900 era surface fires. At the time, foresters did not understand the different frequencies and intensities of fires historically associated with different kinds of forests.

Chief Forester Henry Graves, who succeeded Pinchot in 1910, declared, "The necessity of preventing losses from forest fires . . . is the fundamental obligation of the Forest Service and takes precedence over all other duties."[10] The Forest Service used zealous promotion and propaganda to build an aggressive national program for suppressing forest fires in all forests, including the fire-dependent ponderosa pine forests of the West. People who suggested that fire be used or sometimes be allowed to burn were ignored or put down. Ironically, Gifford Pinchot had suggested in an 1899 *National Geographic* article that the role of fire in creating forests should be studied to help in designing forest management. He wanted to prevent destructive fires but also to understand fire's role as a force of nature. For instance, he recognized that historically fires had helped produce the majestic forests of coastal Douglas-fir, and that without fire they would have eventually been replaced by dense stands of smaller, less-valuable hemlock, due to its ability to grow up beneath and crowd out the fir.[11]

However, learning about the ecological role of fire and applying that knowledge to forest management would take time. The then-embryonic science of ecology would not come of age until the 1960s. Meanwhile there was an immediate need to reduce the threat of damaging wildfires, and by developing a strong program focused on suppression, the fledgling Forest Service solidified its existence and secured increased funding, manpower, and political influence.

The origins of American forestry help explain why it was ill equipped to understand the West's fire-prone forests. Gifford Pinchot and a handful of his contemporaries and colleagues recognized the need for scientifically based forest management and went to Europe to study the relatively new profession of forestry. By the 1890s Pinchot and a few others had begun to apply principles of forestry in some eastern forests, and later they brought them to bear on the western forest reserves. Unfortunately, this proved to be a poor fit.

In Europe, forestry practices were developed to reforest lands that had long ago been denuded by the large rural population to provide timber, charcoal, firewood, and pasture.[12] Foresters established tree plantations on these lands, often introducing new tree species rather than those native to the area. In America, European-inspired forestry was first applied to the humid East Coast to regenerate forests that had been aggressively logged.

In contrast, the West's new forest reserves harbored natural forests, most of which had never been logged. The specific growth and regeneration characteristics of the western forests were adapted to the different frequencies and intensities of fire they experienced. They were the product of fire and other natural disturbances operating over thousands of years. Moreover, most of the drier forests, including ponderosa pine, could not support anywhere near the tree density—the number of trees per acre—that was considered normal in Europe and the eastern United States. In humid regions of Europe, fire in the forest was thought to be completely foreign and destructive. Paradoxically, recent studies of ancient sediments in ponds and other archaeological sites reveal that several thousand years ago, before the development of agrarian populations, most of Europe also supported forests of fire-adapted trees that burned periodically.

Founded on European concepts of forestry, it is easy to see why the US Forest Service came to consider fire in the forest as entirely unnecessary and primarily destructive. This view was enhanced by an out-of-control situation in the West, where people were setting fires at an alarming rate. For instance, after inspecting the Bitterroot Forest Reserve in 1898, John Leiberg reported that "fires kindled by white men have ravaged the forest areas of the reserve in thousands of places." Although the major roads and trails were well posted with a government circular calling attention to the penalty for setting fires, "little attention was paid to it. . . . In two days' travel . . . six camp fires were seen that had been left burning when the campers departed and were slowly eating their way into the adjacent forests."[13]

In addition to fires set by lightning strikes every summer, fires were also started by the burgeoning numbers of homesteaders and other settlers, miners, loggers, railroad crews, and adventurers who had moved into the western forests. They used fire to heat cabins, cook food, and burn debris, all surrounded by forest fuels. Prospectors allegedly burned the forest to expose potential mineral deposits. Railroads spewed sparks from their smokestacks and steel wheels. Fire control was often limited to bucket brigades, if an adequate water source was available, and a few primitive hand tools. Nevertheless, neighbors were often effective in banding together to keep a forest fire away from a home or village, relying on people power and back-burning from roads, trails, or cleared areas to achieve control by removing fuel from the fire's path. The ponderosa forests were not yet overgrown, and thus fires in them could be snuffed out.

In the early twentieth century, the Forest Service concentrated on developing an effective fire-protection organization and strategy. Initially, fires in remote backcountry were simply allowed to burn because of the high costs of putting them out. However, by the 1920s, agency studies showed that it was better to invest more money in rapid initial attack on new fires to achieve early control and thus avoid the much greater expense of fighting large fires. Forest Service policies and procedures were adopted by other federal land management agencies, such as the National Park Service. Forest Service guidance and federal funds were also provided to state fire control agencies, and the US Weather Bureau was recruited to supply fire weather forecasts.

In the early 1930s agency studies continued to indicate that taking prompt, aggressive action on all fires, regardless of location, reduced costs and acreage burned. More roads, better trails, and more fire lookouts and backcountry phone lines were needed to provide access for suppressing fires in the vast backcountry areas of the western national forests. Franklin Roosevelt's New Deal provided the opportunity to make all this happen.

From 1935 to 1937 an abundance of money and manpower through the Civilian Conservation Corps and Works Progress Administration allowed the Forest Service to develop backcountry transportation and communication systems, establish a network of fire lookouts atop ridges and peaks, and build remote guard stations and fuel breaks. Similar improvements were carried out on state lands through cooperative fire prevention programs.

In the early 1930s the Forest Service created a remount depot on the Lolo National Forest in western Montana to quarter and quickly transport pack mules and horses to backcountry trailheads by truck. The mule outfits would ferry fire tools, camp gear, and food to the rapidly dispatched fire crews. These enhancements allowed the Forest Service to adopt a policy ensuring prompt attack on all new fires, even in remote areas. The famous 10 a.m. policy, in effect from 1935 to 1978, specified that the goal of fire crews was to obtain control of any fire before 10 o'clock the next morning. Should that attempt fail, planning and execution would aim to accomplish control before 10 a.m. the following day. Aerial patrols to detect fires became increasingly common starting in the 1930s. Airdrops of supplies followed, and then by the mid-1940s, smokejumpers were parachuting in to

REO Speed Wagon convoy from remount depot transporting mules used to pack supplies to backcountry firefighters (circa 1930s). —*Photo courtesy Archives and Special Collections, Mansfield Library, University of Montana*

fight backcountry fires. The end of World War II brought a wealth of surplus aircraft that became available for aerial fire detection, suppression, and support. By the late 1950s air tankers were employed to drop loads of chemical retardant to quell fires, and helicopters transported specially trained helitack firefighting crews to the scene of new ignitions.

National fire-prevention publicity ramped up during World War II with the threat of Japanese incendiary balloons landing in northwestern forests. Keep Green programs were started in several states and then nationally. The Wartime Council gained permission to use Walt Disney's Bambi and friends on its 1944 fire prevention posters, an effort soon followed by the enormously successful Smokey Bear campaign. The fortuitous and highly publicized discovery of a fire-orphaned bear cub, named Little Smokey, in New Mexico in 1950 further enhanced the fire suppression message. Smokey Bear's message "Only you can prevent forest fires" conveyed to millions of urbanized Americans that fire in the forest was entirely destructive and unnatural.[14]

The great improvements in fire prevention, detection, communication, mobilization, equipment, and other technology seemed to pay off in reducing the extent and damage caused by wildfires in the West. The area burned annually in wildfires in the eleven fire-prone western states had often exceeded 1.5 million acres, from the onset of records in 1916 until the early 1930s. After that period, for fifty years through 1982, no single year ever again recorded that much fire.

Then an astounding reversal took place. Since 1983 the annual extent of western wildfires has *averaged* more than 1.5 million acres, and many years have produced wildfires totaling more than 3 million acres. This modern era has been marked by enormous fires despite continuing improvements in dispatching firefighters, better equipment and technology, and increased expenditures.

Dramatic examples abound. In the summer of 1984, Montana's governor declared an unprecedented wildfire emergency as large, out-of-control fires were burning all across the state's 600-mile width, many of them in ponderosa pine forests. In 1987 the rugged mountains of northern California and southern Oregon were ablaze for two months in the largest fires on record, and a dense pall of smoke filled major valleys, forcing residents to evacuate. In 1988 worldwide media attention was focused on enormous fires in and around Yellowstone National Park. Western forest fires again set acreage records in 1994 and 1996 only to be eclipsed in 2000 and 2002 and then again in 2006, 2007, 2012, and 2013, all involving vast areas of ponderosa pine forest. In 2011, the Wallow fire in ponderosa pine forests

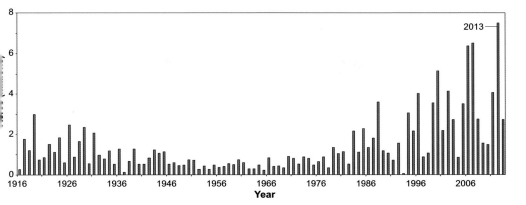

The area burned by wildfires in the West decreased from the 1930s until 1982 because of fire suppression, but then huge fires became fairly common. —*Courtesy Charles McHugh, US Forest Service*

of Arizona and New Mexico set a record for a single fire at 538,000 acres. Federal fire suppression costs have been exceeding $2 billion annually, not counting large expenses for planting trees and measures to reduce damage from postfire erosion. These firefighting expenditures sometimes approach half of the entire Forest Service budget, but they have not helped reverse the West's ever-expanding wildfire problem.

What causes this shocking escalation of dangerous forest fires? There are a number of reasons, from increased tree density to climate. In the last twenty-five years a trend toward warmer than average weather has accelerated the drying of leaf litter, dead branches, and even live trees, making forests more combustible for longer periods. Warmer winters and earlier springs cause mountain snowpack to disappear a few weeks earlier, which further promotes heating and drying.

The increase of trees and dead fuel, a result of fire suppression that kept fire out of most areas for seventy years or more, also contributes to fires that burn hotter, spread more quickly, and are more difficult to put out than those of the past. Visual comparison of hundreds of historical landscape photos taken 75 to 160 years ago with recently taken photos confirms this transformation.[15] Aerial photos that systematically covered national forest lands in the 1930s, compared with recent ones, reveal the same trend: western forests and woodlands have generally become denser. In the past, most forests were either more open or patchy. Forest lands logged a few decades ago now tend to have a thick growth of small trees and sometimes brush. High-elevation forests that were never logged are now covered with older, thicker stands and dead trees.

These two photos were taken from the same photo point in the Lick Creek area of southwestern Montana. The top photo was taken in 1909 after selection cutting. The bottom photo was taken in 1948. A thicket of young fir has filled the open forest, but a few larger ponderosas in the foreground of the 1909 photo are still visible. —*Photos courtesy US Forest Service*

Wildfire suppression policies in place through most of the twentieth century—effective enough in practice to largely exclude fire from forests—are widely blamed for the emergence of today's large, uncontrollable fires. Mark Hudson, in his 2011 book *Fire Management in the American West: Forest Politics and the Rise of Megafires*, argues that the forest products industry was as much to blame as the US Forest Service for these policies. He asserts that those in the forest industry saw fire as a threat to their livelihood and pressured the Forest Service into a policy of fire exclusion. However, Forest Service founder Gifford Pinchot and most early foresters were convinced that open ponderosa pine forests were understocked and producing far below their potential. Foresters also felt that fire scars at the base of trees wasted potential timber. Eliminating fires thus became a primary objective for the first wave of rangers charged with protecting the new national forests in the early 1900s.[16] Limited ability to control fire once it started and failure to understand its critical role in natural ecosystems also helped drive the crusade to eliminate fire from the forest.

Ponderosa forests, which naturally burn every decade or so, have missed more natural fire cycles than most other forest types. Even before the ill-advised fire exclusion policy of the twentieth century, however, there were folks advocating for regular burning of ponderosa forests.

ADVOCATES FOR BURNING

Under the present forest conditions, the no-action or go-slow alternative may very well be the most risky of all.

—Snider, Daugherty, and Wood, "The irrationality of continued
fire suppression," 2006

EARLY ADVOCATES FOR USING FIRE TO TEND FORESTS were a colorful lot. In 1889, Major John Wesley Powell, the one-armed Grand Canyon explorer and head of the US Geological Survey, lectured on the virtues of Indian burning in the forest and described with gusto how he once lit a big forest fire.[1] In 1904, a survey of northern California's Plumas Forest Reserve reported, "The Indians were accustomed to burning the forest over long before the white man came, the object being to improve hunting. . . . The white man has come to think that fire is a part of the forest, and a beneficial part at that. All classes share in this view, and all set fires, sheepmen and cattlemen on the open range, miners, lumbermen, ranchmen, sportsmen, and campers."[2] Indian burning was championed by the embellishment-prone poet Joaquin Miller, who wrote, "By this means [burning], the Indians always kept their forests open, pure, and fruitful, and conflagrations were unknown."[3] A perusal of literature from the late 1800s and early 1900s suggests that some of the fire advocates and early foresters assumed nearly all fires were man-caused and did not realize that lightning was also a major ignition source.[4]

Timber owners liked light burning—setting low-intensity fires in a safe season—as a means of controlling fuel accumulations. Stockmen liked it for stimulating the growth of forage plants. Settlers liked it for land clearing and farming. The frontier laissez-faire burning practices could have coexisted with systematic fire protection, but leading government foresters saw light burning as a political threat, and they refused entreaties from burning advocates to develop procedures for applying fire as a forestry practice.[5]

Early Forest Service studies of fire scars on trees confirmed that a history of frequent low-intensity fires characterized California's magnificent, mixed conifer forests that featured giant ponderosa and sugar pines. But the agency argued that fire scars hastened death and at least lowered the value of trees for lumber. The agency also felt that because fires killed seedlings and saplings, they prevented the forest from becoming fully stocked and producing the maximum quantity of timber. These seemed to be plausible judgments, based on a concept that the West's native forests could eventually be farmed much like forest plantations in Europe. Moreover, the Forest Service reasoned that allowing the use of fire would lead to dangerous blazes.[6]

In the early 1900s as the US Forest Service intensified its program to rid American forests of fires, one region stubbornly resisted. Hot, humid, and sweltering, with an abundance of swampy ground and yet very flammable, the Deep South harbored a rich cultural history of burning the woods, particularly its splendid longleaf pine forests. According to longleaf pine chronicler Lawrence Earley, in early Colonial times a forest dominated by

This ponderosa on the Kaibab National Forest in Arizona has multiple fire scars, visible as vertical black lines on the interior of the cavity.

this stately pine covered 60 million acres along the broad coastal plain from southeastern Virginia to east Texas. Like the West's ponderosa forests, the original longleaf woodlands were mostly open-grown, grassy beneath, and perpetuated by frequent fires. One traveler in Mississippi in 1841 gave this account: "Much of it is covered exclusively with the long leaf pine; not broken, but rolling like waves in the middle of the great ocean. . . . The grass grows 3 feet high. And hill and valley are studded all over with flowers of every hue."[7]

However, by the 1910s when federal forestry began focusing on the South, longleaf pine was not regenerating. Biologists speculated that fire might be important in restoring these pinelands, and the Dean of Yale's School of Forestry, H. H. Chapman, began long-term studies of the effects of fire exclusion and controlled burning in these forests. Excluding fire allowed low brush, palmetto, and other combustible vegetation, known as the southern rough, to build up rapidly. Chapman found that the rough could outcompete pine seedlings, but also that the practice of annual burning to control the rough killed pine seedlings. However, burning at intervals of a few years controlled the rough and allowed the pine seedlings to attain a larger, fire-resistant size. This periodic burning also controlled brown-spot needle disease, which often killed seedlings.[8]

The light burning controversy ramped up considerably in 1910. President Taft, who succeeded Theodore Roosevelt in 1909, appointed Richard Ballinger as Secretary of Interior. Soon Ballinger was accused of virtually giving away federal coal reserves to his industrialist friends. Forest Service Chief Gifford Pinchot, convinced that this was so, publicly denounced Ballinger for corruption. Unable to control Pinchot, President Taft fired him in January 1910, an action that sparked a national controversy because Pinchot was highly respected in the Forest Service and far beyond. The fact that Pinchot's nemesis, Ballinger, supported light burning—stating, "we may find it necessary to revert to the old Indian method of burning over the forests annually at a seasonable period"—certainly didn't help that cause gain favor with foresters.[9] By unhappy coincidence, in August 1910, the same month that the Big Blowup ravaged forests in the Northern Rockies, *Sunset* magazine published an article by timberland owner George Hoxie calling for a government program to conduct light burning throughout California forests. Hoxie argued that we had best adopt fire as our servant; otherwise it will surely become our master.[10]

In October 1910, Chief Forester Henry Graves visited T. B. Walker's extensive timberlands in northeastern California. Graves viewed tracts of ponderosa pine–mixed conifer forest that Walker's crew had methodically burned with a light surface fire after fall rains to reduce hazardous fuel and brush. Graves didn't deny the effectiveness of the treatment but felt it was bad to kill seedlings and saplings. More than that, he didn't like the idea of condoning use of fire in the forest. It didn't help that one of Walker's light burns had blown up earlier in the year and raced across 33,000 acres before submitting to control. Then, like now, deliberate burning in the forest was not risk-free.

Walker and his collaborators tried to persuade the Forest Service to develop light burning into a reliable forestry technique that contributed to wildfire protection. At Graves's suggestion, Walker agreed to underwrite a chair of fire protection at the Yale School of Forestry, contributing $100,000, equivalent to about $2 million in 2014. Graves authorized several field experiments to test the merits of light burning, especially in pine forests of California and the South, but the agency ignored or suppressed any results that might favor burning, while at the same time it promoted its newly articulated policy of comprehensive fire protection.[11]

In the late nineteenth and early twentieth centuries, rural people in the West and South didn't necessarily welcome the intrusion of federal foresters into their lives. Prior to the establishment of the forest reserves, forestland had been largely unregulated, and people grazed livestock, camped, hunted, and harvested trees or firewood for personal use. When the US Forest Service took over management of the reserves, they allowed some public use but under new rules enforced by increasingly visible forest rangers. As time went on, federal and then state authorities began imposing regulations on private landowners to control or restrict hunting, burning, and timber cutting.

Burning the forest was an established tradition, and in 1916 the popular outdoor writer Stewart Edward White lent an articulate voice to the cause of light burning. White owned forest land and argued that burning might be useful for insect control as well as reducing wildfire hazard. Soon, Joseph Kitts, who like George Hoxie was a civil engineer and a California forest owner, joined the burning proponents, writing articles supporting the use of fire to control fuel buildup and opposing the Forest Service's protectionist policy.[12]

The stakes were raised in 1920 when *Sunset* magazine offered an exposé by White on the Forest Service protectionist policy and only agreed to run rebuttals by Chief Forester Graves after the Forest Service threatened to sue.[13] The embattled agency also issued a similar threat to the City of Seattle to remove posters advertising White's position from its public buses. Although the skirmishes between burning advocates in the West and the Forest Service continued through the 1920s, the increasingly influential agency had

Men sawing a large ponderosa pine into shake bolts or firewood with a crosscut saw, circa 1910. —*Photo by Bertie Lord, courtesy Ravalli County Museum and Historical Society photo archive*

by then developed a comprehensive, professionally designed fire protection system, whereas light burners were an ad hoc group that promoted an unproven concept. Only the Forest Service might be able to develop a credible technology for controlled burning, and that agency was adamantly unwilling to do so.[14]

In August 1928, Colonel John White, superintendent of Sequoia National Park, allowed a fire in the park to burn for weeks in defiance of federal and state fire protection policies, and he publicly declared that it had been a beneficial light burn. However, the National Park Service responded to White's challenge by falling into line and appointing a national fire control officer. The 1924 Clarke-McNary Act had provided generous funding for fire protection, dispensed through the Forest Service to federal and state agencies and their other cooperators, and the National Park Service wanted its share. In a sense this could be considered the progenitor of what modern critics have termed the fire-industrial complex, referring to all the fire suppression services, their multifarious suppliers, and other vested interests that depend upon ample congressionally approved funding year after year for all manner of things related to fire control. Historian Pyne concludes that "after 1924 light burning became an official heresy," though it was a concept never completely snuffed out, and like duff smoldering invisibly below the surface, it could burst into flame much later when conditions became favorable.[15]

Studies conducted by the US Bureau of Animal Husbandry showed burning the South's pinelands enhanced their forage value for livestock.[16] Herbert Stoddard, working with the US Biological Survey, published studies in 1931 showing that fire was essential for maintaining habitat for the South's premiere game bird, bobwhite quail.[17] Quail were a critical feature of hunting plantations, an important industry that attracted wealthy visitors. Moreover, the rapid buildup of the southern rough, which was a hazard for uncontrollable wildfires, compelled many field foresters to stubbornly urge Forest Service administrators to allow controlled burning. By 1934, the Forest Service's own Southern Research Station was covertly recommending to administrators that controlled burning be allowed if done for specified objectives by skilled technicians.[18]

Forest Service headquarters in Washington, DC, feared that if it admitted fire could be beneficial in the South and granted permission to burn in southern forests, burning advocates in the West would be emboldened. Thus, the agency continued to suppress and censor findings that supported

Prescribed burning in a mixed conifer forest in Sequoia National Park. —*Photo courtesy National Park Service*

use of fire, as was later revealed in fascinating detail by Ashley Schiff in a 1962 book, *Fire and Water: Scientific Heresy in the Forest Service.* At the same time, the Forest Service covertly allowed controlled burning in the South, sometimes under the guise of "administrative experiments."[19] Finally, in December 1943, the wartime manpower shortage for fighting fires and the swelling tide of evidence and agitation for permission to burn from within and outside forestry caused Chief Forester Lyle Watts to sanction fire's use but only in the South.[20]

Meanwhile, the January 1943 issue of the *Journal of Forestry* contained a startling and revolutionary article by a government forester, making a case for controlled burning in ponderosa pine forests of the West. The disturbing, light burning movement that had been snuffed out by 1930 was suddenly reignited, and for the first time promoted in the foremost forestry journal by an experienced forester. The title "Fire as an Ecological and Silvicultural Factor in the Ponderosa Pine Region of the Pacific Slope" promised that for the first time the case for using fire would be based upon its historical eco- logical role as well as its potential contribution to timber management. The author, Harold Weaver, was employed by the Indian Service (today's Bureau of Indian Affairs) in the Department of the Interior, a relatively little-known agency managing Indian reservation lands, and he built his case based on years of careful observations. Still, as David Carle recounts in *Burning Ques- tions: America's Fight with Nature's Fire*, Weaver's article barely passed through a gauntlet of skeptical reviewers. Also, Weaver's byline in the jour- nal carried the unusual disclaimer, "This article represents the author's views only and is not to be regarded in any way as an expression of the attitude of the Indian Service on the subject discussed,"[21] no doubt in an attempt to shield his employer from Forest Service wrath.

Weaver had graduated in forestry at Oregon State University in 1928, "thoroughly imbued, at that time, with the incompatibility of [ponderosa] pine forestry and fire."[22] Then, as he worked in central Oregon's ponderosa forest, he was shocked when experienced woodsmen and even a renowned forest entomologist—an expert on bark beetles—told him that the policy of excluding fire was a serious mistake. Weaver countered with the standard argument that pines couldn't regenerate if fires were allowed, but the ento- mologist showed him a stand of young pines, many of which had basal scars from having survived past fires. This opened Weaver's eyes. Then, examining young and old pines in many areas, he found they had survived fires at inter- vals mostly between five and twenty-five years. These burns had reduced fuels and thinned young trees, killing more young firs than pines. Inspecting a broad range of forests that were originally dominated by big ponderosas, he found that most had now experienced a long period without fire, and they contained dense thickets of small firs and pines that were often malformed and stagnating.[23]

Disputing conventional wisdom, Weaver's article used observations of tree vigor and other ecological evidence to assert that the thickets of young trees were heavily overstocked and incapable of developing into large trees without thinning by fire or some other means. He pointed out that thinning with fire was more economical than with ax or saw and had the advantage of removing surface fuel as well. Weaver concluded that "converting the virgin [ponderosa] forest to a managed one depends on either replacing fire as a natural silvicultural agent or using it as a silvicultural tool."[24]

Weaver's article was doubtlessly viewed as apostasy by many foresters, although one national forest supervisor congratulated him by saying, "It takes a lot of courage, even in this free country of ours, to advance and support ideas that are contrary to the trend of popular, professional thought."[25] In the years after his groundbreaking article appeared, other foresters who became interested in using fire in ponderosa pine got in touch with Weaver. He conducted burning experiments in ponderosa pine forests of Washington, Oregon, and Arizona, wrote more journal articles, and led field workshops.

Remnants of the historical ponderosa pine forest that grew on this dry site can be found among the stagnant, even-aged pines of today.

Responding to a 1951 article by Weaver, the distinguished University of California forestry professor Emanuel Fritz congratulated him for continuing to study the use of fire in silviculture, adding that "in the early days of forestry we were altogether too dogmatic about fire and never inquired into the influence of fire on shaping the kind of virgin forests we inherited. Now we have to 'eat crow.'"[26]

Weaver's work helped encourage another even more controversial advocate for controlled burning, this time located in California, where light burning promoters had so annoyed the Forest Service. Harold Biswell had earned a PhD in botany and forest ecology at the University of Nebraska. Then he spent several years as a Forest Service researcher in the South, where he became acquainted with controlled burning in pinelands as it was being introduced in the 1940s. In 1947, Biswell became a professor of forestry and plant ecology at the University of California, Berkeley. As he departed the Forest Service, Edward Kotok, chief of research, admonished Biswell to stay out of controlled burning when he got to California. That didn't happen.[27]

In 1945, the California legislature authorized state foresters to issue burning permits for chaparral and other brushlands to improve range and wildlife habitat. Upon arrival, Biswell soon began studying the effects of brushland burning. In the early 1950s he developed a method of firing the bottom of south-facing brushlands in spring under conditions where the fire would die out at the ridgetop when it reached wetter, north-facing slopes. Livestock grazers and wildlife managers liked the results, but forestry authorities became alarmed when Biswell began experimental burning in ponderosa pine forests on the slopes of the Sierra Nevada.[28]

Biswell and Harold Weaver first met in 1951 and then began a long relationship reviewing each other's projects and manuscripts, and as historian David Carle put it, "commiserating with each other's trials." Biswell was introducing controlled burning to large numbers of students, researchers, ranchers, wildlife specialists, and others through his university position, and this outraged some state and federal fire suppression authorities. They demanded that university administrators restrain him, but influential supporters rose to his defense. Biswell persevered, serving a twenty-six-year career at UC Berkeley and, together with Weaver, gaining a cadre of collaborators, adherents, and other allies. Both of these men lived to see the Forest Service make a stunning reversal of policy in the late 1970s and embrace prescribed burning in ponderosa pine as well as in other vegetation types.[29]

While Weaver and Biswell's efforts focused on managed and accessible forests, another area of concern was raised by critics of the fire exclusion policy: a need to return natural fire to wilderness and backcountry. Up until

Prescribed burning to reduce surface fuels and kill small Douglas-fir invading a ponderosa pine forest.

the early 1920s, a few high-level administrators in the Forest Service favored allowing some fires to burn in remote areas based on economic and other practical considerations. Subsequently, however, the Forest Service chose a hard-and-fast policy of completely suppressing all fires. Then in 1934, an unexpected dissenting voice arose from a Montana-born forester who had joined Pinchot's Bureau of Forestry in 1902 and served as a supervisor battling the 1910 Big Blowup. Elers Koch, a well-respected forester in the Forest Service's Northern Region, wrote an essay published in the *Journal of Forestry* that lamented how the complete suppression policy entailed building roads, trails, and phone lines to a network of fire lookouts in the rugged backcountry of north-central Idaho. Koch argued that the area was too rough and erosive for timber management and that forces of nature, including fire, should have been left alone to preserve its special wilderness character. [30] Although the agency's Washington office rebutted Koch's contentions, in a sense it also confirmed them by establishing the 1.9-million-acre Selway-Bitterroot Primitive Area in 1936. Thirty-seven years later the ponderosa pine-dominated canyons of the Selway drainage became the location of the first natural fires deliberately allowed to burn in the Northern Rockies.

In 1924, renowned forester and ecologist Aldo Leopold was instrumental in establishing the first national forest wilderness area, the Gila, in ponderosa pine–covered mountains of southwestern New Mexico. Ever since then ponderosa pine forests have been a focal point for concerns about perpetuating natural ecosystems in the West. Ecologists argued early on that these fire-dependent forests and their big, long-lived trees were jeopardized by the policy of complete fire suppression. This case was presented in conclusive detail by Charles Cooper in his 1960 article, "Changes in vegetation, structure, and growth of southwestern pine forests since white settlement." Cooper concluded that a half-century of fire exclusion was the most important factor in irreversibly disrupting and degrading what had originally been a vast expanse of open-grown, big-tree ponderosa forests.

In the early 1960s, ecological concerns were finally becoming a national issue. A blue-ribbon committee selected by the Secretary of the Interior delivered a groundbreaking report on wildlife management in the national parks that recommended restoring fire as a natural process. The report emphasized that wildlife habitat cannot be preserved in an unchanged condition but instead is dynamic, and that habitat suitable for many species must be renewed by burning. This report helped crack open a door for using fire in national parks and wildlife refuges during the late 1960s and in national forest wilderness during the 1970s.[31]

By the 1970s, most ecologists recognized that natural agents of change like fire, floods, and hurricanes were vitally important disturbances in maintaining natural ecosystems.[32] Today, the concept of returning some form of fire as a process to native forests on public lands has gained scientific credibility. However, a host of economic, legal, political, and logistical constraints stand in the way of reintroducing fire in most ponderosa pine forests, although not so much in large wilderness and backcountry areas.

8

LOGGING LEGACY—FROM CLEARFELLING TO CLEARCUTTING

They attacked the forest as if it were an enemy to be pushed back from the beachheads, driven into the hills, broken into patches, and wiped out.

—Murray Morgan, *The Last Wilderness*, 1955

FIRE WAS THE DOMINANT PROCESS SHAPING PONDEROSA pine forests from the earliest of times until about the mid-nineteenth century. Suddenly a new force appeared in the West's ponderosa forests. Gold rushes in most western states, building of the transcontinental railroads, and booming settlement construction triggered massive logging to provide the wood needed for western development. Over the half century from the late-1840s to the late-1890s, logging of ponderosa pine was a free-for-all, with some loggers high-grading the forest by cutting the biggest and most valuable trees, and others cutting all the trees in a stand with no consideration for the future, an action called clearfelling.

Gilbert Schubert graphically described massive, late-1800s harvesting of ponderosa pine in the Southwest: "Some areas were laid to waste, and huge amounts of slash accumulated which led to some disastrous fires. . . . During the early railroad logging days, large . . . [clearfellings] covered several townships. . . . All failed to regenerate."[1]

Henry Graves, who later succeeded Gifford Pinchot as chief of the Forest Service, surveyed the Black Hills forest in South Dakota in 1896–1897. While clearfelling was commonly used to provide the kinds of material needed for mining, Graves also observed areas where ponderosa pine was being high-grade logged for lumber rather than mine timbers: "The forest after lumbering is composed of large, defective, scrubby trees and small

93

trees not large enough for saw lumber. There are usually as many trees left as have been taken."[2]

Black Hills residents, even in those early years, were alarmed by the plundered look of the logged-over forests that Graves described. In response, general stipulations for cutting ponderosa pine in the Black Hills first timber sale in 1898 required that enough trees remain to perpetuate a productive forest.[3] As the fledgling Forest Service began overseeing larger timber sales in the early 1900s, the agency looked to introduce established forestry methods from the eastern United States and Europe. In 1906, a 37-million-board-foot ponderosa pine timber sale was initiated at Lick Creek on the Bitterroot National Forest in Montana. It was the largest timber sale to date in the newly established national forest system. Elers Koch, a recent graduate from Yale, headed up the young, inexperienced marking crew. In Koch's words:

> Fortunately, ponderosa pine, or yellow pine as it was commonly called, lends itself well to commonsense forestry. One approaches a great yellow-bark pine with the crown beginning to thin out and flatten at the top. No doubt about that fellow; he is mature and ready to cut. . . . Perhaps the next tree is a black-barked tree with a full crown. That fellow is growing fast, and we will leave him for the next cut. So one goes through the stand, taking out the mature trees and leaving the more thrifty to grow. . . . Now we are practicing real forestry.[4]

None other than Gifford Pinchot, the first chief of the Forest Service, inspected the crew's efforts. Pinchot generally approved of their work but recommended they keep a few more large trees that had been marked for cut. Koch visited the Lick Creek area years later and was pleased with what he saw: "The trees have grown enough so the area is ready for a second cut, and all of the openings have seeded up densely to a fine new crop of sapling pine and fir."[5]

The high-grade logging described by Graves in his 1896–1897 survey of the Black Hills and the selection cutting applied by Koch and Pinchot at Lick Creek could both be called partial cutting—some trees were cut and some were left. In the Black Hills, the better trees were removed and the poorer quality trees were left, whereas the opposite was true at Lick Creek. Partial cutting is only a good idea when the best trees remain after logging. Too often in the past, partial cutting has removed the larger, better trees and left the diseased, malformed, and smaller trees. A forest geneticist compared this latter application of partial cutting to shooting the first-, second-, and third-place finishers in a horse race and putting the last-place horse out to stud.[6]

In 1907, an immense, 90-million-board-foot ponderosa pine timber sale was offered on Arizona's Coconino National Forest.[7] The intention of the Forest Service was to gradually convert unmanaged ponderosa pine to managed stands with increased timber production.[8] This philosophy was based on several considerations. First, it was likely a reaction to the earlier widespread use of clearfelling and high-grade cutting. Second, it exemplified the European perspective that mature stands dominated by old, slow-growing trees needed to be converted to young, more productive stands. It also assumed that stands historically kept in open conditions by fire were wasting space that could be used to grow more trees and increase wood production.

Professional management (as opposed to exploitative logging) of the typically uneven-aged southwestern ponderosa pine stands in the early 1900s involved a series of light selection cuttings that collectively removed 60 to 70 percent of the volume and thereby promoted pine regeneration.[9] The importance of securing regeneration could not be overlooked, as ponderosa pine seldom regenerated after the haphazard and heavy logging of the late 1800s, yet it was essential for ensuring a productive forest in the future.

Ponderosa pine stand in 1909 after selection cutting at Lick Creek on the Bitterroot National Forest. Some larger pines were left to provide seed for regenerating openings in the stand.
—*Photo courtesy Archives and Special Collections, Mansfield Library, University of Montana, Missoula*

In 1903, Tim and Michael Riordan, owners of a lumber company in Flagstaff, worried that difficulties in regenerating ponderosa pine could threaten the long-term viability of their business. They contacted their friend Gifford Pinchot for help in finding out why heavily cutover pine stands were not regenerating. Then, as now, connections mattered, and in 1908 the Forest Service established the nation's first forestry research station, now known as Fort Valley Experimental Forest, near Flagstaff.[10]

In 1909, D. M. Lang and S. S. Stewart inventoried the extensive ponderosa pine forests on the Kaibab Plateau north of the Grand Canyon, known as the North Kaibab. Their report suggests that the European approach to management had already influenced their thinking: "The forest under consideration . . . presents a variety of conditions; is so irregular in density, age classes, and quality of timber represented that the usual objects of management must be sidestepped and all efforts directed to securing some sort of silvicultural order."[11]

Lang and Stewart also reported "patchy, uneven aged reproduction" occurring "in scattered clumps . . . varying in age and density." These highly variable conditions did not meet their expectations of a reasonably uniform, professionally managed forest. Managers on the North Kaibab subsequently addressed these conditions using light cuttings, called individual selection cuttings, that removed small volumes of overmature, diseased, deformed, and mistletoe-infected trees.[12] Popular writer John Faris described the process of selecting trees for cutting on the North Kaibab in his signature 1930 book *Roaming the Rockies*: "The plan of the Forest Service [is] to protect the trees, and to limit the cutting of them to those that are ripe for use. These ripe trees are indicated carefully by the authorities. The man appointed to pick out the permitted trees has a special ax bearing the letters U.S.; with this he brands each tree."

Large-scale logging of ponderosa pine started later in the inland Northwest than elsewhere in the West. Geologist Israel Russell explored central Oregon in 1903 and reported that the expansive yellow pine forests were composed of "magnificent, well-grown trees, which will be of great commercial value when railroads . . . bring them within reach of markets."[13] Historian Philip Cogswell later quantified the magnificent ponderosa forest as Russell saw it in 1903: "an estimated 26 billion board feet of it, in open forests on flat or gently sloping ground, waiting, seemingly, for someone to come and cut it."[14]

Getting to the trees was easy—getting them sawn into lumber and transported to major markets was the problem. Central Oregon in the early 1900s was isolated from the rest of the country, and cutting was on hold until a railroad arrived in Bend in 1911. Two Great Lakes lumber companies, Shevlin-Hixon and Brooks-Scanlon, soon moved in and built sawmills on opposite sides of the Deschutes River in Bend.[15] The two companies had been quietly acquiring large tracts of ponderosa forest through a nefarious network of timber claimants known as "dummies."[16] The dummies received all expenses and unknown additional compensation to acquire forested land and then transfer it to the companies. The scheme worked. By the time Shevlin-Hixon opened its sawmill on the Deschutes River in 1916, it had amassed 200,000 acres of prime ponderosa forest.

The companies also brought with them huge skidders that allowed industrial-strength logging. Shevlin-Hixon used a giant, four-line skidder that moved along rail tracks and skidded logs up to 1,000 feet away. A 1918 advertisement featuring Shevlin's four-line machine boasted that it had skidded nearly 52 million board feet of pine the previous year. This capability allowed the two companies to rapidly chew through their forested holdings. In 1924 alone, Shevlin-Hixon processed 200 million board feet of lumber, enough to build about 16,000 small houses.[17] The rapid rate at which the two mills were liquidating their forested ownerships drew attention. Thornton Munger, a research scientist sent to Oregon by Gifford Pinchot, urged the private companies to move away from forest exploitation and into timber farming.[18] Like his counterparts in the Southwest, Munger saw the need to

A Clyde four-line skidder owned by the Shevlin-Hixon Company in a cutover ponderosa forest near Bend, Oregon, circa 1918. —*Photo courtesy Marc Reusser Collection*

develop sound methods for converting slow-growing, old-growth ponderosa forests into fast-growing young forests. In his 1917 treatise *Western Yellow Pine in Oregon*, Munger described the nature of stands at the time: "Yellow pine grows commonly in many-aged stands; i.e., trees of all ages from seedlings to 500-year-old veterans, with every age gradation between, are found in intimate mixture. . . . Usually two or three or more trees of a certain age are found in a small group by themselves."

When it came to managing ponderosa pine, he favored maturity selection cutting: "The system of cutting which seems ideal for this type of forest is a form of selection cutting. Periodic cuttings are made, in each of which all the overmature and thoroughly ripe trees in the stand and all the defective ones are removed; and the saplings, poles, and young thrifty trees are left standing to form the basis for the next crop."[19] Munger was also adamant about "the absolute prevention of forest fires" in the management of yellow pine.[20]

About the same time, Stuart Show and Edward Kotok were analyzing the role of fire in California and were convinced it had no place in managing ponderosa pine forests: "The present California pine forests represent patchy, understocked stands worn down by the attrition of repeated light fires."[21] They did not seem to recognize that these "understocked" conditions had allowed ponderosa forests to sustain themselves for centuries despite frequent fire.

I. V. Anderson, a forester with the Forest Service Regional Office in Missoula, Montana, had a different take. He saw an opportunity to mold management practices to the character of the forest, noting that selection cutting was ideally suited for managing ponderosa pine. In 1933, Anderson observed, "Few timber trees west of the Great Plains are better adapted to selective logging [selection cutting] than ponderosa pine. Fires, insect depredations, and mortality from old age . . . have resulted in uneven-aged stands with a rather irregular distribution of age classes ranging from young seedlings to 600-year-old veterans."[22]

Forest Service scientist Walter Meyer published a report in 1934 on selectively cut ponderosa pine forests in Washington and Oregon. He noted, "The typical ponderosa pine forest of the Pacific Northwest is fairly pure, fairly open, and many-aged."[23] Meyer didn't see many-aged stands as a plus, primarily because small trees were contributing below their potential and too many small trees were now establishing due to the lack of fire.

By the mid-1900s, most foresters experienced in managing ponderosa pine recommended some variation of selection cutting. Notable among these was Gus Pearson, who started working at Fort Valley, Arizona, in 1908 and likely had more knowledge than anyone of southwestern ponderosa pine. Like Munger, he initially favored a form of maturity selection cutting in which the oldest trees were removed first.[24] However, forty years later, Pearson was a strong proponent of improvement selection cutting, a strategy aimed at continually improving the forest by retaining the most vigorous trees of all ages, and recommended light cuttings at about twenty-year intervals.

Region by region then, from the earliest harvesting in the 1840s until the mid-1900s, descriptions of ponderosa pine stands and evolving management practices overwhelmingly recognized that ponderosa forests were fairly open, irregular, uneven aged, and dominated by large trees, with notable exceptions in parts of the Black Hills and Colorado Front Range. From the late 1890s on, a variety of partial cuttings—including high-grade logging, maturity selection cutting, removal of diseased and damaged trees, and improvement selection cutting—dominated the harvest of ponderosa pines and in various ways helped maintain the typical *uneven-aged* stand conditions. And that is why the method introduced to harvest ponderosa pines on the Bitterroot National Forest in the 1960s—clearcutting—jarred the forestry profession and stirred up a previously unengaged public.

Clearcutting is a cutting method that removes all the trees from a tract of forest, or stand, for the purpose of starting a new, *even-aged* stand. It was not a midcourse correction from past management practices in ponderosa pine. Nor was it remotely similar to the brand of selection cutting that former Bitterroot Forest Supervisor Guy Brandborg and his associates had practiced on the Bitterroot National Forest from 1935 to 1955. Brandborg's approach was similar to that of Koch and Pinchot and had received the approval of Pinchot himself during a field trip to the Bitterroot in 1937.[25] The clearcutting approach initiated in the 1960s was a radically different way of managing the forest—completely unlike previous forms of management and inconsistent with the natural processes that historically shaped ponderosa pine forests.

As extreme as the new approach seemed, several developments starting in the 1950s help explain the Bitterroot's decision to begin clearcutting ponderosa pine. Coming out of World War II, demand for lumber skyrocketed with the postwar building boom, and timber from industry-owned lands was becoming scarce. The Forest Service was called upon to increase the volume of timber harvested from national forest lands. Timber output from national forests jumped from 4.4 billion board feet in 1951 to 8.3 billion in 1961 to 11.5 billion in 1970.[26] To achieve these dramatic gains, agency leadership assigned annual timber harvest targets to individual national forests, including the Bitterroot, and the local Forest Service offices strove to meet them.

Sonny LaSalle, a forester involved in the Bitterroot clearcutting controversy, recalls that "rangers were made by their ability to meet timber targets. The pressure was significant."[27]

The business of forestry was also coming of age in the 1950s. Weyerhaeuser Timber Company posted eye-catching advertisements in many popular magazines that portrayed clearcuts with young planted trees and featured birds or animals in the foreground and snow-covered peaks in the distance. The ads tried to show that replacing mature forests by clearcutting and planting was compatible with other forest values, such as wildlife and scenic vistas. Replacing slow-growing natural stands with young, more productive stands had been a long-running theme in forestry, and one that was gaining traction.

About the same time, the American Forest Products Industries (AFPI) launched a radio series aimed at children called *The Adventures of Peter Pine*. On retainer to AFPI, popular Northwest authors Stewart Holbrook and James Stevens (of Paul Bunyan fame) lent their literary talent and credibility to the project.[28] The gist of the story was that Peter Pine was lonely after having lived a long life in the forest, and he wanted to go to town and be with people. After serious thought, a forester went to the woods and marked Peter for cutting. A logger then cut him down and sent him to town, where Peter Pine was converted into useful products for people to enjoy.[29]

The Weyerhaeuser model of clearcutting and planting understandably had appeal to agency foresters, given their need to meet harvest targets. But there was one fly in the ointment, as Sonny LaSalle pointed out: "By law, we had to be able to show regeneration."[30] That was no easy task on the Bitterroot National Forest, given the significant competition for moisture that native grasses pose to pine seedlings. Foresters tried a variety of approaches to increase seedling survival, including scraping a "doughnut" down to mineral soil for several feet around each seedling, but as LaSalle noted, "That was extremely slow and extremely expensive."[31]

Under pressure to get the cut out and then successfully regenerate cutover stands, managers on the Bitterroot were looking for a solution. Axel Lindh, regional chief of timber operations, suggested they consider terracing, which had been used in Idaho and Oregon in the 1950s to regenerate dry, burned-over sites.[32] Terracing was also being used experimentally on the Boise National Forest. After a field trip to examine the practice on the Boise, Bitterroot foresters started employing it on their own forest in 1964. The logic was that clearcutting would maximize the volume of timber removed, while terraces carved by bulldozers would favor planted seedlings by removing competition from other plants, especially pinegrass, and catch moisture.

Terracing also allowed mechanical tree planting, which increased efficiency and seedling survival.

From a technical standpoint, clearcutting followed by terracing and planting successfully accomplished the objective of efficiently harvesting large volumes of timber and establishing new stands of ponderosa pine. However, it fell short based on other measures. It pushed the thin layers of organic matter and topsoil that are important for retaining moisture and nutrients to the outside of the terrace, it was ugly, and it did not emulate any natural disturbance that was ecologically important to ponderosa pine regeneration. It was also costly to build terraces with heavy equipment and then plant nursery-raised seedlings, investment costs that would likely have to be carried 100 years or more. (Bitterroot forest managers at the time calculated 120-year rotations for ponderosa pine.[33]) But perhaps the largest cost to the agency was public trust, because clearcutting and terracing angered a public that had previously assumed foresters were doing the right thing.

The national controversy started out as just a small-time squabble, with a few Bitterroot Valley residents complaining about the in-your-face clearcuts that began popping up in their local national forest. As more people witnessed the new clearcuts and joined the fray, the most formidable opposition

Clearcuts with terraces on the Bitterroot National Forest in Montana, circa 1970.

to clearcutting came from none other than the former forest supervisor, Guy Brandborg. It was a field trip in the summer of 1968 with former district ranger Champ Hannon that really got Brandborg's dander up. After seeing some especially egregious examples of clearcutting and terracing, Brandborg returned home and penned a scathing letter to Montana Senator Lee Metcalf.[34]

Having supervised the Bitterroot National Forest from 1935 to 1955, the silver-haired, plainspoken Brandborg had both credibility and contacts. After several field trips, he was able to engage Dale Burk, a reporter with western Montana's largest newspaper—the *Missoulian*. Burk's hard-hitting nine-part series on clearcutting in the Bitterroot, which appeared in the *Missoulian* during the fall of 1969, really stirred the pot. Brandborg also gained attention from the *Washington Post*, the *New York Times*, and *Reader's Digest*, among others, in exposing the Bitterroot National Forest's clearcutting to the world.[35]

Brandborg was especially upset over what he regarded as blatant disregard for the Multiple Use–Sustained Yield Act of 1960. That legislation states that "the national forests are established and shall be administered for outdoor recreation, range, timber, watershed, and wildlife and fish purposes." In an interview Dale Burk conducted in 1969 for his book *Clearcut Crisis*, Brandborg lamented, "Forestry practices today are entirely different from those applied when I was associated with the Forest Service. I am positively astounded over the scarring, tearing up the landscape, destruction of reproduction and young trees well on their way to provide the next crop of timber. . . Frankly, I don't see how present forest practices in any way meet the requirements of the Multiple Use Act."[36]

Bill Potter, a longtime rancher and timberland owner from the nearby Blackfoot Valley, was equally bewildered why anyone would cut healthy young ponderosa pine, knowing firsthand how hard it was to get them established. He put it as only a rancher could: "First they cut the overstory, then they cut the understory, and that's the whole story. It's like raising a colt until it's big enough to put a saddle on, and then shooting it."[37]

US Senator Lee Metcalf, a Bitterroot Valley native, requested that a committee from the University of Montana's School of Forestry, chaired by Dean Arnold Bolle, conduct an independent investigation of the controversy. In November 1970, the select committee completed its work and presented its findings to the US Senate. This document, informally called the Bolle Report, pulled no punches. The very first item listed under the Statement of Findings echoed Brandborg's earlier supposition: "Multiple use management, in fact, does not exist as the governing principle on the Bitterroot National Forest."[38]

But perhaps the most provocative finding in the report was one the Forest Service least wanted to hear: "The conclusions are clear and incontestable. *Clearcutting and terracing cannot be justified as an investment for producing timber on the Bitterroot National Forest*" (italics theirs).[39]

A simple economic analysis showed that the clearcutting-terracing-planting approach used on the Bitterroot could not come close to breaking even in the long run. The report then went on to answer a critical question:

> If we eliminate timber as a justification for terracing, what is left? Not water. Terracing may not impair water yield or quality, but nobody has shown yet that it improves water production. Not grazing. The purpose of terracing is to eliminate grass and other vegetative competition, which hardly enhances the grazing potential. Not recreation or aesthetics. There seems little doubt that the original forest or a naturally regenerated forest is more pleasing to look at or recreate in. There seems to be no possible way of justifying these practices.[40]

Clearcutting and terracing of the Bitterroot's ponderosa pine forests ended in 1971. The Church Subcommittee Guidelines of 1972 and rules and regulations pursuant to the National Forest Management Act of 1976 substantially limited clearcutting on the national forests. Clearcutting could only be used if it were "determined to be the optimum method . . . to meet the objectives and requirements of the relevant land management plan," and was limited to 40 acres or less in ponderosa pine forests. But the clearcutting issue on the Bitterroot, along with parallel clearcutting controversies on the Monongahela National Forest in West Virginia and the Tongass National Forest in Alaska, energized an awakening environmental movement. As we shall see in the next chapter, it also sparked action on hallmark environmental legislation.

LOVING THE FORESTS TO DEATH

The system that governs the planning and management of the public lands is, or should be, recognized as a snafu decades in the making.

—Jack Ward Thomas, former chief of the US Forest Service, 2002[1]

As THE BACKLASH AGAINST CLEARCUTTING INTENSIFIED, a series of revelations, publications, and events during the 1960s helped launch the environmental movement. Rachael Carson's 1962 book *Silent Spring* documented the dangers of DDT; Lake Erie was declared "dead" from decades of industrial pollution; an oil well blowout off Santa Barbara killed marine life and despoiled beaches along 300 miles of California coastline; and in the Vietnam War, aerial spraying of the jungle defoliant Agent Orange was poisoning people and their environment.

In 1970, twenty million people turned out for Earth Day demonstrations, likely influencing President Nixon to promote a sweeping agenda aimed at protecting the environment.[2] Within a few years Congress had passed the National Environmental Policy Act, Endangered Species Act, and Clean Air Act. Legislation directed at the national forests soon followed, including the Resources Policy Act of 1974 and the National Forest Management Act of 1976. Meanwhile, many Americans were moving from rural areas into the suburbs, but they retained their interest in the outdoors. Visitation to national parks and forests ballooned. A more urbanized society became enthralled by the romanticism of nature and increasingly favored protecting forests from logging and fire.

Although well-intentioned and laudatory in many respects, the environmental legislation hampered ecology-based management and restoration of ponderosa pine forests. The 1970s environmental legislation was designed before there was good awareness of how fire plays an important role in the natural forest ecosystem. The legislation, crafted from a viewpoint that forests should be preserved, reinforced a popular concept in forest ecology that emphasized the climax old-growth forest, the kind of forest that would develop if disturbances such as fires, floods, windstorms, and insect epidemics didn't happen. Paradoxically, the disturbance-free ponderosa pine forest becomes overcrowded with small trees, gradually converts to more shade-tolerant species such as firs in some areas, and transforms from fire- and insect-resistant to being highly vulnerable.[3] A ponderosa pine forest protected from fire would eventually have no large, old ponderosa pines.

Since the 1970s, ecologists have increasingly recognized that disturbances such as fire are vital for sustaining natural ecosystems and promoting biological diversity. Many of our most prized western trees, including giant sequoia, quaking aspen, ponderosa pine, sugar pine, western larch, and coastal Douglas-fir would be eliminated from large areas of their natural habitat without fire, as would dozens of deciduous trees, fruit-bearing shrubs, flowering plants, nutritious grasses, and the animals that depend on these plants. Nevertheless, the environmental legislation of the 1970s promoted keeping fire *out* of western forests.

The legislation requires that land managers prepare environmental impact statements for major proposals, including managing forests with prescribed burns. A Forest Service administrator once said to author Steve Arno, "I would hate to have to write an environmental impact statement that justifies our fire suppression policy." Nonetheless, the policy of putting fire out is thoroughly ingrained in American society and its laws. The Wilderness Act, Endangered Species Act, and clean air and clean water legislation implicitly endorse continued suppression of natural, lightning-caused fires. Environmental regulations also tend to regard prescribed fires that are employed for ecological purposes as impacts on clean air and water rather than as restoration of an important natural process.

And yet protecting the forests from fire in order to protect clean air, water, and endangered species can do more harm than good. The northern spotted owl, a denizen of West Coast old-growth forests, was declared "threatened" under the Endangered Species Act in 1990, putting the brakes on harvesting old-growth forests on federal lands. Biologists thought that protecting large areas of dense, older forests would ensure good owl habitat far into the future. Most owl habitat lies west of the Cascade Range's crest. However, some drier east-side ponderosa pine–fir forests that had grown thick due to fire suppression also hosted spotted owls, and thinning treatments had to be put on hold. Jim Agee, a retired wildland fire expert from

Low-intensity wildfires in open ponderosa pine forests are a natural part of the ecosystem and easily controlled.

the University of Washington, recounts the problem with this hands-off, preservationist approach. In the summer of 1994, he pointed to a large tract of old-growth forest on a map and remarked how vulnerable it was to wildfire. Three weeks later, the 200,000-acre Chelan County wildfires severely burned most of the area, destroying the owl habitat. Agee recognizes the importance of dense, older forests on the landscape; he also notes that "passive management was a dismal failure" for retaining them.[4]

Over time, more high-intensity wildfires destroyed east-side ponderosa pine–fir forests and their spotted owl nest sites—four such fires in 2002 and 2003 on the Sisters Ranger District in central Oregon alone.[5] A host of other animals that inhabit ponderosa pine forests, including the flammulated owl, pileated woodpecker, goshawk, and flying squirrel, have been declared "species of concern." These forests are threatened by severe wildfires, but because of their listed inhabitants, proposed restoration treatments required special handling. Finally in 2011, a revised recovery plan for the northern spotted owl recognized the value of restoration, stating that in habitats east of the Cascades, the only viable conservation strategy will be to actively manage fire-prone forests and landscapes to sustain spotted owl habitat.[6] Active management to mimic natural fires (now largely eliminated) is even proposed by Jerry Franklin and other ecologists for sustaining some of the humid West Coast old-growth ecosystems.[7]

Intense heat from the 2010 Schultz wildfire near Flagstaff charred this ponderosa pine stand and incinerated the organic layer protecting the soil. —*Photo by Brady Smith, US Forest Service, Coconino National Forest*

Continuing to suppress all natural fires requires no special action and largely absolves managers of any wrongdoing, including such things as environmental damage from bulldozing fire-control lines. Even established wilderness areas did not escape impacts of environmental legislation, since it allows suppression of all natural fires. In the largest wilderness areas some natural fires are now allowed to burn under careful monitoring, but most wilderness areas are too small to safely contain fires within their boundaries. If wilderness managers want to use prescribed fire or let lightning-ignited fires burn, they must justify their action in a painstaking environmental review and accept considerable risk if anything goes wrong.

Shortcomings in the environmental laws and regulations might not have been such daunting obstacles to restoring ponderosa pine and other western forests had it not been for loss of trust in federal forestry and the creation of powerful environmental groups. In 1970, when the outcry over clearcuts gained national attention, the Forest Service became defensive.[8] It missed a window of opportunity to restore goodwill and eventually lost the ability to lead in national forest management. Instead, changes in forest management were imposed by Congress and by environmental activists utilizing the courts.[9] During the 1990s the Forest Service began to adopt ecology-based management, but by then flaws in the 1970s environmental legislation, widespread misunderstanding of natural ecosystems, and the power of forest preservationists stifled any transition to ecology-based management. The impasse continues today, and its greatest casualty is the once-magnificent forests dominated by big, old ponderosa pines.

Ironically, an early form of ecosystem-based management, using selection cutting and prescribed burning, had been applied in ponderosa pine forests on Indian reservations since the 1950s by Harold Weaver and others, who espoused these methods in presentations around the West. A handful of Forest Service fire specialists also collaborated with forest managers on scattered ranger districts to apply the methods discreetly in their ponderosa pine forests.[10]

Nevertheless, the top-down direction from Forest Service headquarters was to stay the course and concentrate on producing timber by converting less-productive old-growth forests to fast-growing plantations.[11] The continued emphasis on timber production through the 1980s kept environmental organizations energized. In 1989, critics within the Forest Service established a nonprofit organization called Forest Service Employees for Environmental Ethics, which contributed technical expertise to aid legal challenges against proposed timber sales on the allegation that they violated environmental regulations.[12]

By the 1990s, the kinds of cutting treatments national forest managers designed for ponderosa pine forests would likely have been acceptable to their 1970s critics, but now dozens of environmental nonprofits focused on filing appeals to challenge proposed forest treatments. The timber sales may have had restoration goals, but because they allowed timber companies to cut down mature trees or even to commercially thin younger forests, they were often viewed as bad. The appeal process stopped some poorly planned sales but also delayed projects on overgrown ponderosa forests with hazardous accumulations of fuel.

The nonprofits often made use of the Equal Access to Justice Act—legislation originally intended to help plaintiffs alleging abuse of their civil rights—under which the government pays attorney fees and court costs to plaintiffs that win court cases against federal agencies. A detailed study found that the Forest Service paid out about $1 million per year for plaintiff's costs between 1999 and 2005.[13] The *Missoulian* newspaper concluded that the Equal Access to Justice Act "has become a self-funding mechanism for environmental groups fundamentally opposed to prevailing national forest management direction."[14]

At the same time that the environmental movement was burgeoning and timber sales were being appealed, people were moving into homes and developments located in forests beyond the traditional suburbs. It became feasible to extend power and phone lines into the woods and maybe even to obtain a conventional mortgage loan. Comfortable four-wheel-drive vehicles and pickup-mounted snowplows were now available, allowing year-round access. People bought undeveloped forest land, established a homesite, had a well drilled, and built a house. Although these 1970s forest pioneers were bent on achieving a measure of self-sufficiency—heating with wood, raising a garden, chickens, bees, livestock, and so forth—many still commuted to work in the nearby cities and towns. They were living in the ponderosa pine forest they loved but had access to modern conveniences and amenities.

The ever-expanding zone of residences and related developments located in forested areas near towns is called the wildland-urban interface. Known as the WUI (pronounced "Wooo-eee"), these homes have become problems for land managers and firefighters. Just to be clear, the WUI, broadly defined, overlaps many varieties of forest and shrubland, including California's highly flammable chaparral. However, ponderosa pine is the most abundant forest type associated with the WUI in the West. Ponderosa forest residents are living in a droughty environment with a very long fire season. By casting off resinous pine needles each year that cover the entire forest floor, ponderosas all but provide their own matches. This characteristic benefited ponderosas in the past because it led to frequent burning. Fires stayed on the ground with flames typically just 1 or 2 feet high, thinning out small pines and reducing competition from other conifers. But after a century of fire suppression and logging, these forests are now thick with small trees and massive amounts of woody fuel.

By the late 1980s a few million people had moved into the western forests. For many new residents, their piece of forest was a refuge from the hectic life they experienced when they commuted to work. The new residents often resisted suggestions to thin their own forest, yet most had fire insurance and expected the rural fire department to come to their aid if a fire came near. This wasn't an unrealistic expectation on their part. Wildfires had been successfully suppressed for much of the twentieth century. Television news and movies feature heroic images of brave firefighters and their impressive technology battling destructive fires—men and women in protective gear riding red and yellow pumper trucks into a shroud of smoke and flames, sprawling paramilitary fire camps, valiant soot-covered firefighters wielding heavy tools, smokejumpers parachuting from airplanes, and bombers dumping crimson fire retardant on menacing flames. These superheroes had a record of defeated wildfires, and forest residents felt confident that the fire department could protect them. Few of the new WUI arrivals realized they

The 2000 Cerro Grande fire burned more than two hundred homes within the city of Los Alamos, New Mexico, and the surrounding wildland-urban interface. Note that some homes burned even where trees survived. —*Photo by Andrea Booher, FEMA*

were moving into a tinderbox likely to erupt in a conflagration that destroys trees and houses in spite of the best efforts of firefighters.

In the 1980s and 1990s, many WUI residents paid little more for fire insurance than city dwellers. If they had been required to pay considerably more or to obtain special wildfire insurance, they might have better understood the fire danger. Many mortgages, for example, require people to get flood insurance if they live in the floodplain. But until recently, many insurers were willing to provide coverage for homes in the WUI without first evaluating and rating the fire hazard.

Although city neighborhoods set in ponderosa pine forests are at risk of fire, the *exurbs*—houses dispersed in outlying areas beyond traditional suburbs—are at far greater risk. Many have steep, winding, single-lane access roads threading a narrow passage through a dense growth of trees. They are often poorly mapped and signed. Most lie far from a source of water suitable for refilling a pumper truck. At a recent conference on wildfire, one speaker referred to these homes as "exurban woody fuels." These outlying residential areas are estimated to account for at least half of the annual $3 billion fire protection cost borne by federal taxpayers in recent years.[15]

Exurban housing in ponderosa pine forests threatens wildlands like national forests, national parks, and other preserves, because it prevents managers from using prescribed burning or allowing natural fires to burn to maintain the ecosystem. Managers cannot take the chance that a prescribed fire will escape and burn down houses. Managers who are trying to restore the forests must also deal with the endless complaints about smoke from prescribed fires, noise from equipment such as logging trucks, and degraded views. Homesites have often been as effective as spotted owl nest sites at preventing restoration treatments on public lands.

The population of ponderosa pine country has grown rapidly. Eight States that make up most of the interior West more than quadrupled in population between 1950 and 2010, nearly twice the growth rate of the United States as a whole. A large portion of this increase is in ponderosa pine forests, much of it near the boundaries of undeveloped public lands maintained for wildlife habitat and other ecological values. An analysis published by the National Academy of Sciences concludes that the main threat to these protected forest areas is housing growth in the adjacent privately owned land.[16] We are loving our forests to death.

FORESTS UNDER SIEGE—FROM MEGAFIRES TO BARK BEETLES

I'm afraid that the future of our forests is going to be more fires like Las Conchas and Wallow. We have a goddamn mess on our hands.

—Bill Armstrong, fire specialist, Santa Fe National Forest[1]

IN THE LATE 1940S, EMINENT ECOLOGIST Aldo Leopold warned that overzealous efforts to make forests safe were unnatural and would have dire consequences. "A measure of success in this is all well enough," he wrote, "but too much safety seems to yield only danger in the long run."[2] Those words sounded strange if not heretical at the time, but decades later, Leopold's prescient warnings became reality. By the 1980s, a new kind of wildfire—the megafire—was blazing through the West's ponderosa pine forests. Megafires kill trees across tens of thousands of acres, feature flame lengths hundreds of feet high, incinerate homes by the tens or hundreds, and cost taxpayers millions of dollars. Megafires are crown fires of unprecedented magnitude—leaping from tree crown to tree crown and killing nearly all trees in their wake. We use the terms *crown fire* and *stand-replacement fire* interchangeably to mean fires that kill trees, although it should be noted that under certain conditions surface (ground) fires can also kill overstory trees.

Fire has been part of ponderosa pine forests for millennia, although massive crown fires have not. Much of the landscape prior to Euro-American settlement was covered with dry grass and pine needles, allowing fires to occasionally burn and expand for months before being checked by late fall or winter weather. Research by Pete Fulé and his colleagues at Northern Arizona University suggests that wildfires as large as 60,000 acres occurred in major fire years, such as 1785, a year when many old ponderosas in the

Southwest were scarred by fire.[3] Such large fires were likely a combination of low-severity surface fires in ponderosa pine forests that gradually transitioned into high-severity crown fires as they spread upward into higher-elevation mixed conifer forests.

Studies of historical fires in ponderosa forests across the West corroborates that prior to 1900, pine forests experienced mainly surface fires. Crown fires were found to play a more important role in the Mt. Rushmore area of South Dakota, but even there accounted for less than 4 percent of the area burned.[4] Stand-replacing fires have also been documented in some ponderosa forests occupying steep terrain along the Payette River in Idaho.[5] However, the most prominent historical role of mixed- to high-severity fire appears to be in higher-elevation ponderosa pine forests along the northern Front Range in Colorado.[6] Evidence of stand-replacement fires has also been documented in ponderosa forests atop sky island mountain ranges in the Southwest, but these fires were typically limited to patches of less than 200 acres.[7]

Crown fires were seldom mentioned by early-day observers of ponderosa pine forests, and then only on a much smaller scale than seen today. Charles Cooper, who studied changes in southwestern pine forests from the

The flaming front of the Rodeo fire on June 19, 2002, as it surged over the Mogollon Rim west of Show Low, Arizona.
—*Photo by Jerry Beddow, US Forest Service Air Attack*

time of Euro-American settlement up to 1960, reported that despite the high frequency of surface fires, stand-replacing fires were rare.[8] Cooper searched the early literature and failed to turn up a single report of a crown fire in Arizona before 1900. The first large, stand-replacement fires were the 1951 Escudilla fire, which burned about 19,000 acres, and the 21,000-acre Dudley Lake fire of 1956. Ironically, the area burned by the Escudilla fire was reburned in 2011 by the Wallow fire, which was more than twenty times larger.

The nature of early fire in California ponderosa pine forests was evidently similar to that in the Southwest. In 1897, George Sudworth described conditions on the Stanislaus and Lake Tahoe Forest Reserves: "The fires of today are peculiarly of a surface nature, and there is no reason to believe that any other type of fire has occurred here. . . . The instances where large timber has been killed outright by surface fires are comparatively rare. Two cases only were found. One of these burns involved less than an acre, and the other included several hundred acres. They are exceptional cases."[9]

The slow transition from historically dominant surface fires in ponderosa forests to more intense crown fires is likely the reason it failed to create alarm until the late 1970s, when previously unusual crown fires were gradually becoming the new normal. High tree densities and continuous forest canopies over large areas were fueling ever-larger crown fires and, in dry years, sometimes megafires.

Drought is an overriding factor associated with major wildfire years. Detailed tree-ring analysis has allowed scientists to establish fire-climate correlations in ponderosa pine forests dating back several centuries. For example, Thomas Swetnam and Christopher Baisan at the University of Arizona studied a network of sixty-three sites across the Southwest and found that all high, or bad, fire years in ponderosa forests occurred under severe drought conditions.[10] While not all drought years were high fire years, all high fire years were drought years. Whether fires in drought periods become large and intense depends on whether they encounter dense forest canopies and heavy surface fuels. The recent spate of megafires suggests those conditions are commonly met today. If the trend toward earlier springs and warmer than average summers persists, as current science suggests,[11] the era of megafires can be expected to continue and worsen.[12] The story of two notable megafires—one in 2002 and one in 2011—demonstrates the firefighting challenges and devastating effects of this new breed of fire.

The spring of 2002 was powder dry in the Southwest and followed a winter with little precipitation. By June, 100 percent of Arizona was rated in severe to exceptional drought.[13] The parched conditions put fire management staff on edge as the spring wore on. Kate Klein, then ranger on the Black Mesa Ranger District in eastern Arizona, recalls how the relentless drought, low humidity, and wind combined to set up the perfect firestorm. "I could feel it in the air," she recalls, "just a sense of tension."[14] Managers hoped they could hold out until the traditional summer rains showed up in July. But luck was not with them. On the afternoon of June 18, 2002, an out-of-work firefighter started a blaze at the Fort Apache rodeo grounds near Cibecue in hopes of creating a demand for his services. He didn't anticipate that the fire he ignited, aptly named the Rodeo fire, would soon provide employment for nearly 4,500 firefighters. Two days later, a lost hiker intentionally set another fire about 20 miles away—near Chediski Peak—in an attempt to signal a helicopter of her whereabouts. The woman was rescued, but crews did not reach the fire before it raced up the mountain to become the Chediski fire. On June 23, the two fires merged, and news of the Rodeo-Chediski fire soon changed from a report on the radio to a massive 30,000-foot plume visible for miles. *Helplessness* was the word firefighter Paul Summerfelt used to describe the gut-wrenching feeling that overcame him at the sight of the plume: "You could just watch the cloud of smoke on the horizon. You knew it was coming and you knew it was bigger than you were going to be able to deal with."[15]

Jim Youtz, now a silviculturist at the US Forest Service regional head-quarters in Albuquerque, was a Bureau of Indian Affairs forester on the Fort Apache Reservation at the time. While he acknowledges wind was a factor, he describes the Rodeo-Chediski as "primarily a fuel- and drought-driven fire."[16] Former Forest Service district ranger Jim Paxon, who served as the information officer and national spokesman during the course of the fire, agrees: "It was pretty much fuels-related, fed by the millions of excess trees in our overcrowded forests. It had extremely high energy. When I started fighting fire in the late '60s we didn't have these big columns or plumes that would build up, collapse in an explosion on the ground, and create hurricane winds."[17]

Paxon later wrote a book, *The Monster Reared His Ugly Head: The Story of the Rodeo-Chediski Fire,* that recounts the day-by-day unfolding of the fire through his eyes. Paxon says the combined fire was unlike any he or other firefighting veterans had ever seen. It was "off the fire-behavior calculation charts," and could not be fought. He describes a wall of orange 6 miles wide as it came over the Mogollon Rim, throwing flames 400 feet into the air—at one point burning 10,000 acres in fifteen minutes.

The Rodeo-Chediski's intense heat charred about half the area within the fire's perimeter. Timber losses on the Fort Apache Reservation alone were estimated at $300 million. The Rodeo-Chediski fire also burned 465 homes and 26 commercial buildings and outbuildings. The 468,000 acres burned in the Rodeo-Chediski fire appeared to set an unreachable record at the time, but it was a record that would not stand long.

Less than a decade later, in 2011, two careless campers abandoned their campfire in the Apache-Sitgreaves National Forests, and it later grew into the largest wildfire in southwestern history—the 538,000-acre Wallow fire. Jim Zornes, deputy forest supervisor on the Apache-Sitgreaves, described conditions when the Wallow took off: "We knew we were in extreme conditions. We had fuel everywhere and our probability of ignition for any fire that hit the ground was 100 percent. With 62-mile-per-hour wind gusts, it was blowing so hard it was tough to walk."[18] Zornes says he'll forever remember first seeing smoke from the Wallow fire, then 30 miles away. "We knew we had a tiger by the tail, but we didn't know how big." Once he realized the fire's horrific potential, he says "I just almost fainted."[19]

The Wallow and Rodeo-Chediski fires shared several characteristics. Both fires burned through large areas with fuel conditions far outside historical norms. Both fires also included smaller treated areas of open, uneven-aged ponderosa pine that survived relatively intact and now provide a valuable reference as managers develop treatments for ponderosa pine restoration projects.[20]

The Rodeo-Chediski and Wallow fires also differed in important ways. The Rodeo-Chediski mainly burned through dense timber in steep terrain. The Wallow fire was wind-driven and burned across a rolling landscape of forests intermixed with meadows and cinder cones. Virtually all of the area within its perimeter burned, but only about one-quarter of the area burned severely.[21] Large areas burned at low intensity, producing desirable ecological effects similar to prescribed burning. However, the Wallow fire, too, left its mark. Thirty-five homes and six outbuildings burned, a number that would have been much higher except that large areas adjacent to the communities of Greer, Nutrioso, and Alpine had been thinned in the previous few years.[22] But Paul Summerfelt, veteran firefighter and now wildland fire management officer for the City of Flagstaff, sees this as a somewhat Pyrrhic victory: "We herald the success of protecting towns like Greer. Firefighters were able to make a stand in the treatment areas when seemingly all was lost. But what does that mean long-term when you look out the window and what you see is black? And the forest, except for the immediate vicinity of the community, is gone, and they're facing potential flooding issues, and tourism decline, and all of those damaging ecosystem effects that remain after the fire is out and the smoke is gone."[23]

Residents anxiously watch the menacing plume of the rapidly approaching Rodeo-Chediski wildfire. —*Photo courtesy* Arizona Daily Star

Homes burned by the Rodeo-Chediski fire. —*Photo courtesy Humphrey's Type 1 USDA/ USDI Southwest Region Incident Management Team*

Megafires—An Autopsy

Indeed, it is only after the smoke is gone that forest managers and wildlife biologists can evaluate the damage. Impacts from recent megafires include destruction of wildlife habitat, degraded watersheds, and soil damage so severe in intensely burned areas that new trees will not grow back in any reasonable time frame. More than a decade after the Hayman fire charred nearly 140,000 acres southwest of Denver, a 50,000-acre patch near the center of the blaze remains treeless. The Hayman fire cooked this core area as though it were in a giant barbecue, and all ponderosa pine seed sources were destroyed. Only a few thousand acres replanted since the fire are beginning to hint at a new ponderosa pine forest. It will likely remain grassland and shrubland for decades to centuries thinks Peter Brown, a forest ecologist and director of Rocky Mountain Tree-Ring Research in Fort Collins.[24] His work shows that much smaller patches of ponderosa forest left bare by earlier crown fires have not regenerated, so he holds little hope for the huge area scorched by the Hayman fire.

Megafire impacts on threatened and endangered species have been especially dramatic. "The Mexican spotted owl is the biggest concern we have as an Endangered Species that we're trying to help," says Jim Paxon, veteran of the Rodeo-Chediski fire and now spokesman with the Arizona Game and Fish Department. "The Forest Service is under extreme pressure not to do any cutting around the nesting sites." And that irony makes the aftereffects of the 2002 Rodeo-Chediski fire and the 2011 Wallow fire all the more frustrating to him. "Between the two fires, we lost 20 percent of the Mexican spotted owl nests that exist in the world" he says.[25] The unprecedented damage from the two megafires served as a catalyst for revising the federal Mexican spotted owl recovery plan, which now identifies large, uncharacteristic wildfire as the primary threat to the future of the owl.[26]

Stephanie Coleman, fisheries biologist on the Apache-Sitgreaves National Forests, documented the effects of the Wallow fire on the Apache and Gila trout, which are listed as threatened species. The Fish Creek system, which consists of three streams and one lake, was an Apache trout recovery system on the Apache-Sitgreaves. The system was stocked in 2006–2007 with Apache trout and comprised the largest population of the species on the forests.[27] Postfire flooding in Fish Creek reached 1,000 cubic feet per second, washing away all of the riparian area, including the gabion barrier that kept nonnative fish species from entering the stream. Coleman reports that postflood fish surveys found only one remaining Apache trout in Fish Creek, which is now considered a total loss for the population and habitat.

Raspberry Creek, also on the Apache-Sitgreaves National Forests, was a recovery stream for Gila trout. Although burn severity within the Raspberry Creek drainage was not especially high, Coleman notes that postfire flooding removed the entire population and greatly altered in-stream habitat by filling in pools with sediment. "We now have no Gila trout populations left on the forest," she says.[28]

Prior to the fire, the Arizona chapter of Trout Unlimited had focused its efforts on stream and riparian-area restoration. After observing recent megafire impacts on recovering trout populations, the organization has shifted its emphasis toward restoration of the surrounding forests. The fires have also brought about a change in thinking among wildlife biologists in general—from a focus on protecting certain species and their habitat to a broader interest in landscape-scale forest management and restoration.

Megafire effects on forests and wildlife, particularly threatened and endangered species and their habitat, make headlines, but the economic and social costs associated with megafires are also alarming. The federal government spends huge amounts of taxpayer dollars annually to protect millions of homes in the wildland-urban interface, and yet damages from megafires continue to mount.[29]

A detailed study of wildfire costs was published by the Western Forestry Leadership Coalition in 2010. Surprisingly, suppression costs, which are the costs associated with putting a fire out, were a minor part of overall costs for most fires profiled.[30] Cost data from other sources also show that total wildfire costs are not necessarily highly correlated with the size of fire, although fires larger than 50,000 acres are almost always expensive. Small wildfires, such as the 2007 Angora fire near Lake Tahoe, California, could also be considered megafires if costs are considered. Suppression costs alone for this 3,100-acre fire came to over $11 million. Total costs, including suppression costs, compensation for property losses, and rehabilitation efforts, came to $160 million.[31] Much of the Angora fire was in a forested residential area, or wildland-urban interface (WUI). Proximity of the fire to Lake Tahoe and concerns about sedimentation from runoff brought intensive rehabilitation efforts, which included applying mulch by helicopter, constructing sediment barriers, and reseeding native plants. Even given the complexity of fighting WUI fires like the Angora, observers such as Roger Kennedy, former director of the National Park Service, are skeptical of skyrocketing costs. "There is a fire-industrial complex," he says. "A lot of people are making money on this."[32] With increased use of airpower and well-equipped ground crews, fighting fires costs a lot of money. At the height of the Angora fire, managers deployed 186 engines, 24 helicopters, 15 water tenders, and 2,180 personnel.[33] Skeptics blame a "costs be damned, do whatever it takes" approach to fighting fire for suppression costs that exceeded $3,500 per acre on the Angora fire.

The June 2010 Schultz fire in Arizona, which covered a modest 15,000 acres, is another blaze that would not meet conventional megafire standards based on size. However, this fire burned hot enough to kill virtually all trees over broad areas of the San Francisco Mountains north of Flagstaff. Steep, burned-over areas had no green trees or forest litter remaining to buffer torrential rains and hold the soil in place. Monsoon rains in July and August of 2010 scoured unprotected slopes and sent huge volumes of water and sediment into the valley below, drowning a young girl. Such intensely burned and disturbed soils also favor invasion by exotic plant species.[34] Rory Steinke, watershed specialist on the Coconino National Forest, headed up the mitigation efforts on more than 5,000 acres of the burn. Steinke says that costs of sowing native plant seed, mulching, controlling invasive weeds, and reclaiming the area as a functioning watershed have tallied over $4 million so far. Several million more will likely be needed to stabilize stream

Torrential flooding that followed the 2010 Schultz wildfire near Flagstaff resulted in massive erosion and debris movement. —*Photo by Ann Youberg, Arizona Geological Survey*

channels.[35] Despite these huge expenditures, mitigation efforts can only reduce impacts—not undo the enormous damage done. One bright spot in the otherwise dark cloud that was the Schultz fire came in 2012, when residents of Flagstaff overwhelmingly passed a $10 million general obligation bond that funds forest treatments around the city to reduce threats of wildfire and damage to the municipal watershed.[36]

Bark Beetles

Megafires in ponderosa pine forests have grabbed headlines during recent years, yet bark beetles have killed millions of ponderosas over a similar period with far less fanfare. One such example comes from a ponderosa pine forest adjacent to an Arizona megafire. In July 2002, just after the smoke had cleared from the Rodeo-Chediski wildfire, author Carl Fiedler observed a lot for sale just outside the fire's perimeter. If the absentee landowner had called the realtor to determine the status of his property, the realtor may have responded, "I have good news and bad news. The good news is that the fire stopped just short of your lot. The bad news is that most of your big pines are dead. The fire missed your property, but the bark beetles didn't."

Bark beetles killed these large ponderosa pines, which were located just outside the perimeter of the 2002 Rodeo-Chediski fire.

Beetles can sense easy pickings. Stressed ponderosa pines emit volatile terpenes, which allow bark beetles to detect trees vulnerable to attack. Beetles bore into the inner bark (phloem) of the tree and then chew tunnels, or galleries, in the phloem that serve to girdle the tree. Attacking beetles emit an aggregation pheromone that attracts other beetles, and collectively they overwhelm the tree. Bark beetles also carry blue stain fungi, which invade the sapwood of the tree and inhibit movement of water from the soil to the needles, causing the needles to turn brown.

Three species of bark beetles account for most of the damage in ponderosa pine forests: mountain pine beetles, western pine beetles, and engraver beetles. The mountain and western pine beetles belong to the genus *Dendroctonus*—Latin for "tree-killer"—a fitting descriptor for these two aggressive, tree-killing insects. Engraver beetles are generally less aggressive but may cause significant tree mortality, especially during periods of drought. Several other bark beetles, including the roundheaded pine beetle, can also be lethal.

Historically, bark beetles and fire served as natural tree predators in ponderosa pine forests.[37] Both agents "thinned the herd," with surface fires killing many small trees and beetles sporadically killing individual trees or clumps of medium-sized and larger trees. Together, fire and bark beetles helped create the open-grown, uneven-aged forests of the past. Beetle outbreaks occurred occasionally in denser patches of forest, particularly during drought periods when lack of moisture stressed the trees. Scientists theorize that localized beetle outbreaks ebbed and flowed historically as dense forest areas and drought conditions came and went, but region-wide epidemics were likely less common than observed today. Perhaps the best-documented early bark beetle epidemic in ponderosa forests occurred in South Dakota's Black Hills. This epidemic started in 1895, and by 1902 entomologist Andrew Hopkins reported that "the Black Hills beetle [mountain pine beetle] has killed most of the trees from Deadwood to the Wyoming border."[38] During the course of the epidemic, which lasted until 1908, an estimated 10 million trees were killed.[39] Six more beetle epidemics irrupted in the Black Hills over the next century, with the most recent one starting in 2005.

The scale of historical mountain pine beetle epidemics in ponderosa pine, although occasionally large, pales in comparison to the 1930s western pine beetle epidemic that raged through old-growth ponderosa forests in the Pacific Northwest. This outbreak killed nearly as much ponderosa pine volume from 1931–1935 as was cut for timber, and fifteen times the amount destroyed by fire during the same period. By 1943, after the epidemic had run its course, the western pine beetle had killed nearly 16 billion board feet of ponderosa pine in Oregon and Washington alone.[40] The western pine beetle completes up to four generations per year; hence, populations can explode under favorable conditions.[41]

The most recent bark beetle outbreak started in 2000, and by 2003 a full-blown epidemic was underway in both ponderosa pine and higher-elevation lodgepole and whitebark pine forests. The epidemic peaked in 2009 and has since declined due to more normal weather patterns and because many of the vulnerable pines have already been killed. Overall, 23.6 million acres of forest have been impacted by the mountain pine beetle in the western United States; when all bark beetles are included, that number more than doubles to 47.6 million acres. Much of the acreage impacted by beetles is occupied by lodgepole pine, but because ponderosa pine is a component of nearly two-thirds of all forests in the western United States, impacts in many ponderosa pine forests have also been severe.[42]

Western pine beetle galleries etched into an old-growth ponderosa pine in Oregon. *—Photo by US Forest Service, 1926, courtesy Forest History Society*

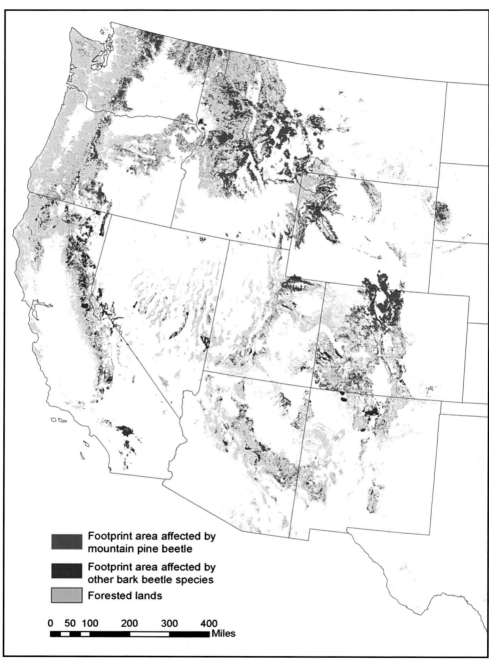

Map of the western United States showing areas affected by bark beetles for the period 2000–2011. —*Map courtesy Chris Fettig, US Forest Service*

Legend:

- Footprint area affected by mountain pine beetle
- Footprint area affected by other bark beetle species
- Forested lands

0 50 100   200   300   400
Miles

The current beetle epidemic differs from historical ones in several ways. Today's dense, mostly even-aged ponderosa pine and mixed pine–fir stands regenerated after heavy logging a century or so ago. While some of these stands have been thinned or partially cut, many have not. Crowded, unthinned stands are especially vulnerable because the maturing trees are stressed and unable to produce enough pitch to ward off invading beetles. Recent warmer, drier conditions that favor beetle population growth also render ponderosa pines more vulnerable to attack. Historically, beetle outbreaks were less severe in ponderosa than in lodgepole pine forests because of ponderosa pine's enhanced capability to "pitch out" attacking beetles, and because ponderosa forests were more open and contained trees that varied considerably in size, age, and vigor, and therefore in susceptibility.[43] In some parts of the ponderosa's range, the mountain pine beetle is secondary to the western pine beetle as a tree killer, such as in large-tree-dominated pine stands in California and the ponderosa pine belt of the Pacific Northwest.[44]

The geographical range of bark beetles is limited by climate, a characteristic that likely explains the recent expansion of mountain pine beetles into previously beetle-free areas.[45] Recent collections in Nebraska document the presence of mountain pine beetle in that state for the first time, and mountain pine beetle populations have also been found in Alberta in areas not considered part of the insect's historical distribution.

Both megafires and bark beetle epidemics are triggered or intensified by drought and overcrowded forest conditions. With several million acres of ponderosa forests blackened by megafires since 2000, and bark beetles killing trees on additional millions of acres, the ponderosa pine faces an uncertain future. In response to this dire situation, forest managers are developing innovative, landscape-scale projects for restoring sustainable ponderosa pine forests.

RESTORATION—IS IT TOO LATE?

Nature in the twenty-first century will be a nature that we make; the question is the degree to which this molding will be intentional or unintentional, desirable or undesirable.

—Daniel Botkin, *Discordant Harmonies*, 1990

CHRIS PILESKI, A STATE FORESTER IN MILES CITY, described recent wildfire impacts on the rolling ponderosa pine forests of southeast Montana: "The Whitetail Divide area used to be some beautiful country with scenic views and big trees. . . . Now it's all black. Get up in a helicopter near Colstrip and as far as you can see, it's burnt—pretty depressing." He estimates 160,000 acres of ponderosa forest burned in 2012 alone—80 to 85 percent in stand-destroying fire. "I've heard people say that if we keep up this pace, all the forest in southeast Montana will be burned in twenty years. I can't say that I disagree."[1]

Anne Bradley, longtime ecologist with the Nature Conservancy in New Mexico, laments that recent megafires have left huge swaths of the Southwest devoid of ponderosa pine. The lack of surviving ponderosas, coupled with continuing drought, portend that large areas will remain treeless. Ominously, the megafire phenomenon seems here to stay, with increasingly overcrowded forests, more people living in the woods and thus more ignition sources, and projections of a warmer, drier climate. Wildfire isn't the only threat, though. A recent study in the Southwest relating tree mortality to drought foresees possible conversion of some forests to shrublands or grass by midcentury.[2] Stressed ponderosa forests also provide raw material for unprecedented insect outbreaks, as evidenced by the recent bark beetle epidemic across the entire West. The confluence of these major threats inevitably leads to the question: Are efforts to restore ponderosa pine too late? To address this question, in this chapter we present a primer on restoration and examine its potential to rejuvenate and sustain at-risk ponderosa forests.

Beginning in the 1990s, national forest managers were in a bind. They saw the growing threat from severe crown fires, but because of public reaction to heavy cutting of the previous decades, they were unable to gain support for treating large forested areas commensurate with the scope of the problem. They were able to implement less controversial treatments, including light thinnings that only removed some of the sapling- and pole-size trees on small areas of the forest, and heavier thinnings and fuel treatments along primary roads. These latter treatments, called forested fuel breaks, created open stands of uniformly spaced trees. Forested fuel breaks reduce the volume and continuity of fuels and create what firefighters call "defensible zones" that enhance the opportunity to suppress fires.

A special Forest Recovery Act passed by Congress in 1998 allowed creation of more than 180,000 acres of forested fuel breaks along travel routes located on the eastern slope of the Sierra Nevada. The effort will eventually encompass most of the 1.5 million acres of east-slope national forestlands

Forested fuel break along a highway in northern California. In addition to heavy thinning of the live trees, woody fuels on the ground are burned to reduce fire hazard. Note the denser forest in the background where the thinning ended. —*Photo by Chuck Lewis, US Forest Service*

The Camp 32 fire calmed down when it entered a thinned forest on the Kootenai National Forest in northwestern Montana. The fire burned from the bottom of the photo, crossed the road in the middle of the photo, and then dropped to the ground and did not kill the trees in the thinned forest in the upper half of the photo. —*Photo courtesy US Forest Service*

north of the Lake Tahoe basin. A 2010 study showed that these treatments have effectively moderated wildfire behavior and resulted in a wide range of ecological and social benefits when compared to untreated areas.[3]

In August 2005, the Camp 32 fire blew up in an untreated ponderosa pine forest and headed toward an area of forest homesites near Eureka in northwestern Montana. Then the fire hit a 600-acre tract of the Kootenai National Forest that had recently been thinned and subjected to prescribed burning. The flames immediately dwindled to no more than knee height and were easily controlled.[4]

Since the 1990s, several million acres of ponderosa pine forests have either been lightly thinned or thinned more heavily around developments or along roads to create fuel breaks. Still, this represents only a small proportion of the land originally resplendent with the pine forests extolled by early visitors. Furthermore, light thinnings alone are not sufficient to allow large trees to develop or to substantially reduce crown fire hazard. Fuel break treatments are effective in reducing crown fire hazard but do so by creating open stands with uniform size and spacing of trees. Both of these treatments have their place, but neither is designed to address the full range of problems in declining ponderosa forests. However, a nascent discipline called restoration was also taking root during the 1980s and 1990s.

Beginning in 1984, a study comparing cutting and burning treatments with an adjacent uncut forest was installed on the University of Montana's Lubrecht Forest and replicated on a nearby guest ranch. This study was controversial because it featured selection cutting (which retains some trees across the full range of diameters) and prescribed burning, and it focused on the trees to leave rather than the trees to cut—all radical concepts in forestry at the time. However, differences among these treatments twenty-five years later were dramatic: On the uncut plots, dozens of old-growth ponderosas had been killed by bark beetles, the remaining live pines were stressed and growing slowly, and Douglas-fir were taking over in the understory. Ponderosas on the same-size plots receiving selection cutting and burning were vigorous and growing rapidly, and few large trees had died. Initial five-year results from this study, combined with prescribed fire research conducted by the Missoula Fire Sciences Laboratory in the 1980s, led to a joint Fire Lab–University of Montana restoration study at Lick Creek on the Bitterroot National Forest in 1991. This was the same area on which Elers Koch and Gifford Pinchot had implemented a form of selection cutting in 1906, and the same national forest that had sparked a firestorm in the late 1960s because of clearcutting. Results from Lubrecht, Lick Creek, and subsequent demonstration projects have helped shape the approach to ponderosa pine restoration in the Northern Rocky Mountains.[5]

In the mid-1990s, Wallace Covington and a team of scientists at Northern Arizona University (NAU) began intensively studying ponderosa pine forests in the Southwest. Their efforts led to the founding of the Ecological Restoration Institute (ERI) a few years later. The beautiful southwestern pine forests so admired by visitors and residents alike were not only declared unhealthy by ERI scientists but also at great risk. Their conclusion was that ponderosa forests needed "restoration," and the best way to do that was by removing substantial numbers of trees followed by prescribed burning.[6] This was an eye-opening and controversial revelation, given the region's low level of timber harvest and general disenchantment with tree cutting. The ERI documented vast differences between historical and present-day forest conditions and identified numerous benefits following restoration treatments in overcrowded ponderosa forests. Their investigations went far beyond the typical forestry research of the time; restoration treatments were found to increase resistance to crown fires, promote productivity of understory plants, and even enhance butterfly populations.[7]

In 1996, scientists at the Blacks Mountain Experimental Forest in northern California began comparing cutting treatments that maximized structural diversity (leaving trees of many sizes at irregular spacing) with those

View of an untreated plot in 2009, twenty-five years after the Lubrecht Forest–guest ranch study started. Note the large fallen tree in the foreground and recently killed old-growth ponderosas with rust-colored foliage in the background.

View of an adjacent treated plot in 2009, twenty-five years after selection cutting and prescribed burning.

that minimized diversity (leaving trees of the same size at uniform spacing), with and without prescribed burning.[8] And just a few years later, managers from the surrounding Eagle Lake Ranger District of the Lassen National Forest began implementing restoration treatments as part of their day-to-day management.

Despite these noteworthy early efforts, restoration is a new enough discipline that there is not yet a settled definition among those who practice it. Nature writer Gary Nabhan calls those who work in this field "ecosystem physicians,"[9] while retired University of Montana professor Jim Habeck thinks of restoration as "repairing a damaged ecosystem."[10] Restoration implies putting something back the way it was—a fairly straightforward task when it involves a nonliving object like an old car. However this becomes a more challenging proposition for a forest ecosystem.

Most people now agree we must restore pine forests—but to what condition? We can't precisely duplicate the historical forest because the influences that shaped it, such as insects, fire, climate, and land use were unique to a given time period and location. However, we can use knowledge of historical ponderosa pine forests and the processes that helped shape them as a guide for treatment. Painstaking research has reconstructed the size, number, and arrangement of trees in numerous historical forests, which profile a range of sustainable forest conditions in the past.[11] Contemporary forestry research has also established that open forest conditions are needed to increase vigor in old-growth trees, reduce crown fire hazard, and establish healthy pine regeneration.

How this restoration work gets done varies by region. In the Northern Rockies where second-growth stands dominate, the preferred treatment approach only looks to approximate characteristics of sustainable forest conditions, including low-density large trees—along with trees of many sizes, if available, and random to clumpy arrangement.[12] In ponderosa forests of the Southwest, which typically contain some old-growth trees, the preferred approach is quite specific. It looks to recreate stand conditions that closely resemble those at the time of Euro-American settlement (circa 1880) in terms of the size, number, and location of the remaining trees.[13] Both approaches, as well as the one at Blacks Mountain, recommend follow-up prescribed burning. They also focus on the trees that are kept—that is, the future forest—rather than the trees that are cut. Marking the trees to keep helps ensure that the healthiest trees are reserved, and it provides a visual impression of the number, size, and spatial arrangement of the trees that will remain after treatment.

Selection cutting and thinning are especially useful tools for reducing tree density wherever they can be employed, because unlike burning, they

Historical photos help document forest conditions of the past. Upper Yosemite Valley viewed from Columbia Point in Yosemite National Park in 1866 (top) and in 1961 (bottom). Note how much denser the forest was in 1961, nearly a century later. —*1866 photo by H. G. Peabody, courtesy Robert Gibbens; 1961 photo by Robert Gibbens*

allow choosing which trees are removed and which are left. To restore the forest to a less vulnerable condition, many trees need to be removed. Today's ponderosa forests commonly have ten to twenty times more trees per acre—mostly small and medium-size ones—than they did prior to 1900. Because these forests are stressed due to prolonged overcrowding, fire is often not the best tool for thinning trees at the first stage of restoration. Many of the excess trees are too big to be killed by low-intensity fires. And any blaze of sufficient intensity to kill unwanted medium-size trees would weaken and sometimes kill large trees and invite bark beetle attack. However, prescribed burning is a superior tool for reducing surface fuels and killing excess seedlings and saplings. Prescribed burning in combination with thinning has repeatedly made the critical difference in reducing damage from extreme wildfire events, in contrast to the 2006 Tripod Complex fire in Washington, where thinning alone did not prevent a crown fire.[14] Fire is also a key disturbance agent for rejuvenating understory plants. Numerous forage plants—especially shrubs—need to be top-killed occasionally to sprout vigorously, and seeds of some plants lie dormant in the soil for years or decades until fire induces their germination.

Although selection cutting or thinning are the safest restoration treatments, fire may be the only treatment allowable in national parks and wilderness areas. Recent research indicates that burning alone may be an effective treatment in some situations. However, this approach requires a series of burning treatments to kill both small and some medium-size trees and to consume the fuel resulting from earlier fire-killed trees whose scorched needles, trunks, and branches accumulate on the ground. Multiple burning treatments are costly and often difficult to schedule due to limited periods of favorable conditions and a shortage of burning crews. They also come with potential legal liabilities, and heavy smoke or escaped fire can trigger public criticism.

Where managing for timber products is a primary objective, such as on state-owned school trust lands, industrial timberlands, and some private and tribal lands, a modified restoration plan for ponderosa pine forests might approximate current management, which is to culture and maintain moderately open stands. Chipping the slash or burning it in piles might be used in lieu of prescribed fire.

Restoring the few remaining old-growth forests of ponderosa pine represents a special challenge. The centuries-old pines are typically weakened due to an unprecedented increase in younger trees. Shade-tolerant firs and even young ponderosas in these stands have developed extensive, shallow root systems as a result of fire suppression, giving them a competitive advantage over the old growth in extracting soil moisture and nutrients. These

Litter (needles and exfoliated bark) gradually builds up around the base of a tree and then burns during a fire. The depth of litter and duff removed by this prescribed fire did not seriously injure the large ponderosa pine.

sapling-, pole-, and sometimes medium-sized trees also predispose the old-growth to death from wildfire and bark beetle outbreaks. Removing most of the competing smaller trees may be necessary for stressed old-growth trees to recover and survive.[15] People who manage ponderosa pine forests in the West need to have a long view, since the trees far outlive humans.

The initial restoration treatment should be followed by a maintenance treatment in as few as ten years for burning, or twenty to thirty years for selection cutting or thinning. Because the second treatment usually removes fewer trees and generates less slash, it offers a better opportunity to burn some of the accumulated litter and duff throughout the stand. Where prescribed burning isn't feasible, slash can be piled and burned in openings. Pile burning temporarily reduces competing ground cover, making the openings ideal for pine regeneration, whether natural or planted.

Despite a compelling case for restoration, it won't happen unless the costs are bearable and the benefits substantial. If there is no marketable use for the trees, the initial thinning and removal of excess trees can cost thousands of dollars per acre. Large volumes of trees and slash also pose a major disposal problem. Burning heavy volumes of this material in piles can sterilize the underlying soil, and limited landfill space cannot be spared for this potentially useful organic material.

This ancient ponderosa (top), growing within a dense stand of younger pines near Boulder, Colorado, was highly susceptible to crown fire. The old pine, whose basal scars document that it had survived numerous surface fires over its estimated 600-year lifetime, was killed in the 2010 Fourmile Canyon wildfire (bottom). The dead tree was cut down because it was deemed a potential hazard to hikers. —*Photos by Priscilla Stuckey*

Healthy ponderosa pine saplings in a forest opening. These fast-growing trees can easily be aged by counting the number of annual branch whorls.

These spindly little trees grow near those in the above photo but have stagnated after growing years in shady conditions. Such suppressed trees—often thirty to forty years old—can only be accurately aged by cutting them at ground line and counting their rings under a microscope.

People who want to restore the forest must view the wood products industry as a partner. Selling harvested trees for products can help pay workers to thin the forest and remove hazardous woody fuels. A primary obstacle to restoration is that many pine forests needing treatment occur in areas with only limited wood-products infrastructure. With no market for the many tons of small trees that need to be removed on a single acre of overgrown forest, costs for disposal are prohibitive, and experienced and efficient contractors are nowhere to be found. The 2008 Great Recession set back efforts to restore forests because the market for wood products crashed, many mills went out of business, and loggers and others in the wood industry abandoned their trade.

Jerry Williams, retired director of fire management for the Forest Service, makes the point that there is no silver bullet for restoration. In his view, "Restoration needs to take a comprehensive approach where economic incentives, new markets, education, legislative changes, and other inducements must all work together in accelerating restoration work. If any one element is missing, the whole effort seems at risk. An early assessment analysis[16] indicated the need to treat three million acres per year in the eleven western states, giving priority to [ponderosa pine and associated species] and [to] stick with it for a minimum of fifteen years. . . . We've been stuck at about 300,000 acres per year for nearly two decades . . . one-tenth [of] the need . . . while high-intensity wildfires continue to consume vast areas."[17]

Despite some encouraging developments, there is an inherent reluctance in government and among private landowners to invest in managing forests to reduce fire hazard. A study reported in the *Economist* found that thirty times more is spent on responding to disasters than on preventing them.[18] Accomplishing restoration at the scale that is needed won't be cheap, but neither is fighting megafires. And mitigating post-megafire ecological damage is not restoration—it is picking up the pieces.

A big step in scaling up restoration efforts came in the form of the 2009 Collaborative Forest Landscape Restoration Act. This federal legislation provided the policy framework and funding for landscape-level projects on federal lands. By 2014, over a dozen large-scale restoration projects in ponderosa pine forests had been funded. Here we examine the potential of two such projects, one in Arizona and another in Oregon, to turn the tide in ponderosa forests.

Restoration Meets Collaboration

The Four Forest Restoration Initiative (4FRI), or Four-Fry as it's called, is a massive 2.4-million-acre landscape restoration effort just getting off the ground in Arizona's Apache-Sitgreaves, Coconino, Kaibab, and Tonto National Forests, the largest contiguous ponderosa forest in the West. The impetus for such a large project started much earlier, however. Step back to the year 2000, and the Forest Service in the Southwest was in a pickle. On one hand, the few remaining sawmills were pressuring the agency to make more timber available for harvest. On the other, environmental organizations were trying to keep the Forest Service out of the woods. Mike Williams, forest supervisor on the Kaibab, recalls the gist of a discussion shortly after he arrived in 2001: "We have to start thinking big; we need to be treating landscapes rather than stands."[19] Go back a few years more, and Kate Klein, then district ranger on the Apache-Sitgreaves Black Mesa Ranger District, recalls beginning what would be years of stymied attempts to implement a large-scale fire hazard reduction project. In 1996, district personnel began evaluating the potential for thinning 30,000 acres just southwest of the communities of Heber and Overgaard in east-central Arizona. Managers were concerned that if a wildfire started in the dense ponderosa pine forests, prevailing southwest winds might fan a firestorm that could overrun the two towns.

Many people with homes or cabins in the area were retirees or vacationers, and when asked about the proposed thinning project "most of them opposed it, because they moved here for the forest" says Klein.[20] But by the fall of 1999, the district had developed a modified plan to treat only one-third of the area. However, several environmental groups appealed the project, contending it did not adequately assess effects on either the environment or the endangered Mexican spotted owl.[21]

After the appeal was denied by regional Forest Service officials in February 2000, the groups filed a lawsuit to stop the project, sending Klein and her staff back to the drawing board. After much additional work, the final Forest Service rebuttals to the lawsuit were sent to the regional office on June 17, 2002. On June 18, the Rodeo fire broke loose, and two days later the Chediski fire erupted. "Four or five days later the whole area had burned up," Klein says, along with 426 homes and 55 Mexican spotted owl Protected Activity Centers.[22] "Talk about frustration and hopelessness and anger and depression!"[23]

The 468,000-acre Rodeo-Chediski fire—the largest and most destructive wildfire in Arizona up to that time—turned out to be a primary catalyst for taking action and thinking big. Phrases like "game changer," "wake-up call," and "eye opener," are used to describe the Rodeo-Chediski fire and its aftershocks. 4FRI team leader for the Forest Service, Henry Provencio, calls the fire a "turning point." Dick Fleishman, assistant team leader of 4FRI, says that after the Rodeo-Chediski fire "the freak-out factor got real large, real fast." Jim Youtz, regional silviculturist with the Forest Service in Albuquerque, simply says, "It got the public's attention."

The finger-pointing that followed the Rodeo-Chediski fire resembled a circular firing squad. After emotions died down, there was general recognition that business-as-usual was detrimental to everyone's interests. The Arizona Governor's Forest Health Council was convened in 2003 to develop a strategy for addressing the state's hazardous forest conditions. In 2003, the US Congress passed the Healthy Forest Restoration Act (HFRA), which focused on reducing wildfire hazard and encouraged collaboration. The HFRA was accompanied by a categorical exclusion that limited public comment on projects smaller than 1,000 acres, a feature viewed by some as a means of fast-tracking timber harvests and limiting opportunities for environmental review. Rick Tholen, a recently retired forester who was the Bureau of Land Management's HFRA coordinator for several years, believes the law tended to alienate the various interest groups rather than bring them together.[24] Despite this and other legislative attempts to expedite restoration treatments, things were taking too long and too little area was getting treated. While the various factions were fiddling, the Southwest's pine forests were burning up.

In late 2008, a diverse cast of thirty entities, including the Arizona Game and Fish Department, Center for Biological Diversity, Flagstaff Fire Department, Gila County, Grand Canyon Trust, Northern Arizona Logging Association, Rocky Mountain Elk Foundation, and the Nature Conservancy, began meeting as a stakeholders group. Not long after, Congress passed the Collaborative Forest Landscape Restoration Act of 2009, which provided the administrative framework and financial support to get a large project off the ground.

The Forest Service selected Henry Provencio, a biologist on the Coconino National Forest and thirty-year Forest Service veteran, as the agency representative to the 4FRI stakeholders group. When Provencio first heard of the immense scale of 4FRI, he said, "That's crazy, there is no way in hell they're going to get this done." But after reading the in-house solicitation for the team-leader position and thinking about the positive ecological implications, he says, "The landscape scale made all the sense in the world."[25]

Provencio recalls his first meeting with the stakeholders group early in 2010, where he made it clear the 4FRI project was for real, and their collective responsibility was to make sure it kept moving. "It's like a train," he said, "You can either help steer, shovel coal, or get out of the way."[26] Dick Fleishman, also an Arizona native and Forest Service veteran, is the assistant team leader. Fleishman, always ready with a pithy observation, demonstrates a broad and deep knowledge of the agency's inner workings and protocols. Together, the experienced duo projects a quiet confidence that they are up to the task of collaborating with stakeholders, navigating the project through the agency's regulatory web, and getting it implemented.

The sheer scale of 4FRI also requires a mind stretch. Bill Noble, wildlife biologist on the project, was at first overwhelmed by the huge data requirements for 4FRI compared to conventional projects. "It's ridiculous," he says, "but then I realized that ridiculous was what was required."[27] The Forest Service has issued nineteen task orders for restoration treatments under the current 4FRI phase 1 contract, with an additional 20,000 acres under other contracts. A draft Record of Decision was published in December 2014, which, when final, will give statutory clearance for about 430,000 acres of restoration cutting treatments. Given this approval, Fleishman expects about 20,000 acres (2015) and 30,000 acres (2016) of 4FRI landscape will be treated under the combined 4FRI phase 1 and timber sales contracts.

Jim Youtz, US Forest Service regional silviculturist, says that characteristics of previously treated stands that survived the Rodeo-Chediski and Wallow fires were an important reference for designing 4FRI restoration treatments. Forest Service staff also used reconstructions of historical (pre-1880) ponderosa pine stand conditions that were maintained under frequent low-intensity fires,[28] and input from the stakeholders group, as primary sources for designing treatments. Rather than simply thinning to reduce fire hazard, 4FRI used these sources to design cuttings that create a diversity of tree age classes and historical tree patterns in the forest. Neil McCusker, 4FRI silviculturist, uses the term "spots on a dog" to describe what the pattern of trees and openings in treated areas might look like from the air.[29]

Virtually all areas receiving cutting treatments in 4FRI will also receive prescribed burning, and many additional acres will receive burning only. Overall, approximately 60,000 acres per year will need to be burned. Given that an average of 20,000 acres per year burn in wildfires, an average of 40,000 acres will need to be treated annually with prescribed fire.[30] 4FRI fire ecologist Mary Lata sees that as a tall order, especially in drought years and years with major wildfires in the region. Over time, Lata says, one benefit of such an aggressive burning program is that the burning windows, the conditions under which burning can be accomplished, will widen, because less fuel will accumulate, allowing prescribed fire to be conducted under drier conditions.[31] 4FRI is aiming to reburn areas at roughly ten-year intervals, Lata says, because fuels build up too much if burning occurs at longer intervals. The problem is that at year eleven and beyond, fire managers would face not only the usual 60,000 acres of first-time burning, but an equally large acreage needing reburning, whether by prescribed fire or wildfire.

Fire creeps through the needles and grass during a prescribed burn on the Coconino National Forest, Arizona. —*Photo courtesy Mary Lata, US Forest Service*

In addition to 4FRI's challenging burning schedule, Kaibab National Forest public affairs officer Jackie Banks worries that stakeholders may suffer burnout, noting that several members of the initial planning group have dropped out.[32] The group has been meeting regularly since 2009, and often has to work through contentious issues before reaching a consensus. Despite the obstacles, the project has reenergized some national forest staff. Provencio says that one late-career employee on a ranger district even told him, "This is the first time in a long time that I've been excited about a project."[33] Commissioners of counties within 4FRI boundaries are also pleased that the associated jobs will help boost their local economies and relieved that their communities will gradually become safer from wildfire as treatments are implemented. The large size of 4FRI also makes possible, and actually requires, reestablishment of a wood-processing industry, which will greatly increase the opportunity to utilize large quantities of small trees across the region and thereby benefit non-4FRI forest thinning projects as well. How 4FRI plays out in the coming years will be a measure of the potential for collaborative, landscape-scale restoration throughout the West.

Long before the 4FRI project went from concept to reality, a landscape-scale restoration effort was taking form in the Lakeview region of south-central Oregon, locally known as Oregon's Outback. It's so remote that the US Census Bureau designates the area as "frontier" rather than "rural." Timber and cattle have largely supported the hardy residents of this high, dry country since the town of Lakeview was established in 1876. The climate is well-suited for ponderosa pine, and the area's impressive pine forests were a natural draw for early-day logging and settlement. Concerns about sustaining timber's critical role in the local economy date back to 1950, when the Lakeview Federal Sustained Yield Unit was created on the Fremont National Forest.[34] The purpose of the sustained yield unit was to reserve timber harvested within its boundaries for manufacture by mills in the communities of Lakeview and Paisley, thereby helping stabilize their economies by eliminating outside competition.[35]

But despite the apparent advantage of the sustained yield unit designation, things were not going well for Lakeview in the early 1990s. Paul Harlan, vice president of resources for Collins Pine Company in Lakeview, says there just wasn't enough timber available to keep Lake County's five sawmills in business. The situation deteriorated further following federal agency rule changes in 1994 that reduced harvests in Pacific Northwest forests. Although the 1994 Northwest Forest Plan only applied to national forests west of the Cascade Range, the associated controversy spilled over to the east side. Fearing lawsuits to block timber sales, the Forest Service issued interim guidelines that same year protecting old-growth forests east of the Cascades—called the Eastside Screens. The new guidelines banned cutting trees larger than 21 inches in diameter.

Harlan saw this as a death knell. The Forest Service "by edict, shut the forests down" he says.[36] By 1996, only two mills remained in Lake County, and the one in Paisley closed that year. It was a devastating event for such a small community, Harlan recalls, and "almost everyone was there to watch the last log go through."[37] Locals were angry and resentful and blamed environmentalists.

After a period of mourning the mill's closure, Jane O'Keeffe, a fourth-generation rancher and Lake County commissioner, and Paul Harlan, manager of the only remaining sawmill, decided to take action to keep Lakeview and Lake County from dying a slow death. O'Keeffe hit the road to drum up awareness of the town's plight. She traveled far and wide, knocking on doors and inviting people to come to Lakeview to see firsthand what was going on. Many of the folks O'Keeffe invited were seen as adversaries at the time—representatives from a bevy of environmental organizations in Portland, Eugene, Seattle, and Washington DC. Her efforts garnered funding support from Sustainable Northwest, a Portland-based nonprofit organization, to sponsor a meeting in Lakeview during the summer of 1998.[38]

O'Keeffe and Harlan didn't know what to expect as some ninety people showed up for the meeting, including civic leaders, residents, loggers, scientists, environmentalists, and businesspeople. Harlan says that a meeting in the same place several years earlier had deteriorated into a shouting match, and he didn't want a sequel. "I wanted people to leave their guns at the door and talk about what's happening out on the forest," Harlan recalls.[39] As it turned out, all parties behaved themselves as O'Keeffe and Harlan recounted the heavy collateral damage to the community resulting from the Eastside Screens and years of forest gridlock.

Harlan made it clear that they were not looking to defend the old ways of logging. "It's a new day and a new way," he said. The meeting was followed by a day in the woods. Enough common ground was reached to warrant a follow-up meeting, and then another, and thus the Lakeview Stewardship Group was born. The group, which includes representatives from west-side environmental organizations, has continued to meet since 1998, making it one of the longest-running of its kind in the country.

Since its designation as a federal sustained yield unit in 1950, harvesting in the Lakeview Unit had focused on removing the valuable larger trees, mainly ponderosa pine. However, in keeping with Harlan's pledge for change, the newly organized Lakeview Stewardship Group developed a restoration-based approach for managing the unit, and in 2001, the Sustained Yield Unit was redesignated as the Lakeview Federal Stewardship Unit.

A long-term goal of the collaborative group's restoration plan for the unit is to reduce wildfire hazard and gradually increase percentages of ponderosa pine in forests where it is being replaced by white fir and lodgepole pine. A top priority is reinvigorating the approximately 200,000 acres of old-growth ponderosa pine that occur in the unit, much of it in deteriorating condition. Restoration treatments are typically aimed at removing the fir, lodgepole pine, and juniper that have invaded the old-growth stands. In younger, second-growth stands, cutting treatments preferentially remove fir and lodgepole to increase water retention, reduce risk of insects and disease, and create sufficient space between trees to reduce wildfire hazard and promote regeneration of ponderosa pine.

In 2008, Collins Pine entered into a ten-year stewardship contract with the Forest Service to conduct timber harvesting and restoration treatments in the unit. Proceeds from the sale of timber products pay for restoration activities, including such things as reclaiming unneeded forest roads, restoring willow and aspen in riparian areas, removing invasive plants, and clearing debris from tributary streams to allow free passage for fish.

Mike Anderson of the Wilderness Society says what started out in Lakeview as a cautious experiment has turned into a success story.[40] He says that as the group has evolved, issues can be resolved in less and less time. Another longtime participant, Rick Brown with the Defenders of Wildlife, mentioned that he used to think it was all about the forest. But his thinking has changed. "It's about forests *and* people," says Brown.[41] Indeed, it is the potential impact on people in and around Lakeview that distinguishes this landscape-scale effort. It is a homegrown project, and nearly everyone in the community is affected by it in some way.

Stewardship group member Craig Bienz, an ecologist with the Nature Conservancy, says the Lakeview project aims to implement restoration treatments (cutting and prescribed burning) on 15,000 to 20,000 acres per year.[42] Trees removed in restoration treatments provide about one-third of the Lakeview mill's timber needs, which helps keep the mill running and loggers and mill workers employed. Lakeview District Ranger Amanda McAdams says the biggest challenge to the success of the Lakeview project is finding a way to handle the mountain of small trees that is generated annually by restoration treatments.[43] Burning piles of small trees in the woods is doable but problematic due to high cost and air quality concerns, and they cannot be left untreated because of wildfire concerns.

The ongoing threat of wildfire poses another problem. Bienz points out that the 2012 Barry Point wildfire burned 43,000 acres within the Lakeview Stewardship Unit, including two project areas that were scheduled for treatment in 2013–2014.[44] The dilemma is that the funding request the group submitted to the Collaborative Forest Landscape Restoration Program did not anticipate dealing with project areas that burned before they could be treated. The group is now deliberating the salvage of trees killed by wildfire, which must be utilized within twelve to eighteen months to still have value for sawn products. The salvage cutting will not advance the Stewardship Group's restoration objectives, yet the trees are needed to provide raw material for the mill and are essential to meet the economic and employment objectives of the project. Regardless of the outcome, the Lakeview experience should alert other collaborative groups to propose options for addressing this kind of dilemma before it occurs.

Collaborative projects, such as the ones in Arizona and Oregon, offer the best and perhaps only hope for getting restoration work done at a meaningful scale in a contentious world. A fundamental difference in restoration projects of just a few years ago and those funded under the Collaborative Forest Landscape Restoration Program is that the latter are being implemented at a much larger scale, involving thousands of acres. These projects focus on overall restoration needs—not just fire hazard reduction—and include treatments aimed at controlling invasive plants, removing unused roads, enhancing wildlife habitat, and improving watersheds by planting willows and replacing clogged or undersized culverts. Large-scale projects also help sustain existing sawmills, and in situations like the 4FRI project in Arizona, may provide the critical additional wood supply necessary for new, state-of-the-art wood-processing facilities to be built. Collaboration may be the bridge to a different, more effective way to accomplish restoration in the future, or with refinements it may actually be the future.

The importance of finding a process to implement landscape-scale restoration projects cannot be overemphasized. Restoration treatments have been field tested and have proven successful.[45] The challenge now is to find a way to get them on the ground rapidly and over large enough areas to ensure a healthy future for ponderosa pine. However, while an aggressive restoration program is being carried out on national forests and other federal lands, it is critical that Americans who live within the forest do their part also.

PROTECTING A
HOME AND ITS FOREST

*Firefighters told John they thought they had the fire
out, but his shake roof caught fire again and the house
burned to the ground.*

—*Spokesman-Review*, "Hangman Hills
lessons largely ignored," 1991

IN 2000, THE CERRO GRANDE FIRE CHARGED INTO the forested city of Los Alamos, New Mexico. The stifling smoke, hazardous fuels, and huge number of homes in the fire's path overwhelmed firefighters. Although the fire did not race through tree crowns, it spotted ahead by way of airborne firebrands. More than two hundred houses burned, primarily those with wooden shake roofs or adjacent flammable landscaping, pine needle litter, wooden decks, and woodpiles. Most houses without these outside combustibles were untouched, and most of the ponderosa pine trees not close to a torched house survived. The nearby Los Alamos National Laboratory facilities were spared, thanks to a thinning and fuel removal project completed a few years earlier.[1]

On several recent occasions, a raging wildfire entered a tract of forest that had previously been thinned and the slash and understory fuels burned or removed. In each case, the towering wall of flames suddenly died down and flames no longer charged through tree canopies. Fire intensity nosedived, and only low flames traveled across the ground. In 2010, the Fish Hatchery and Slide Creek fires torched ponderosa pine forests in northeastern Washington. Homeowners in the fires' paths had thinned and cleaned up fuel on their properties, which helped firefighters protect fifteen homes and thirty-three outbuildings. When the fires entered the thinned forest, they dropped out of the tree canopies and burned more moderately along the ground, allowing most trees to survive.[2]

In July 1994, central Washington's Tyee Creek fire devastated 140,000 acres including nineteen homes. Flames were running through the ponderosa pine–Douglas-fir forest canopy as they approached Christine and Mike Mallon's isolated ranch house in a mountain canyon. Between 1989 and 1991 under a cooperative agreement, the Mallons had thinned and piled slash on several acres of national forest land immediately adjacent to their home, and the agency had burned the piles. As the crown fire swept through, it dropped to the ground in the fuel-treated stand, and both the trees and the Mallon home survived.[3]

On May 6, 2009, a man-caused fire flared up on a hot, windy day in the drought-plagued forest at Timberon in New Mexico's Sacramento Mountains. The crown fire raced through the dispersed community, reaching Don and Ruby Roberts's property within minutes. For years, the Roberts had

A wildfire in a ponderosa pine–fir forest destroyed this home and killed the surrounding trees. Despite the presence of a pond and small lawn, combustibles around the house led to its incineration.

thinned their ponderosa pines, removed many understory trees, and pruned the tall trees near their house. The raging fire and the volunteer firefighters arrived at the same time. Reaching the treated forest, the fire calmed down, allowing the Roberts's home and many of their trees to survive.[4]

While it helps if the government is actively promoting these efforts, as an individual you can make a big impact by restoring the forest around your home. Your neighbors will see the benefits of your stewardship in the beauty of your carefully thinned forest. When they ask how you did it or why you did it, you can mentor them about the purpose and benefits of forest restoration. By becoming an advocate, you can also aid stewardship of nearby public lands. Managers of these forestlands are often swayed by local sentiment; it is easier to thin forests adjacent to private property whose owners favor restoration than to thin forests next to unwelcoming landowners. Forest homeowners who demonstrate good stewardship on their own property have credibility with land managers, other landowners, and the general public.

Today, a great many residents of ponderosa pine forests have not accomplished basic self-protection—that is, making their homes and surrounding property fire resistant. Consequently, every year many homes in these forests burn despite massive expenditures for firefighting. Montana Governor Brian Schweitzer sternly warned people living in the wildland-urban interface (WUI) that if they want their homes to survive a wildfire, they need to take personal responsibility and create a firewise homesite: "Don't look for the government to bail you out."[5] When flames are racing toward homes in the woods, firefighters will be there to help, but in the face of perilous, flaming fronts, priority will be on evacuating people and saving lives—not on hazardously situated buildings. Moreover, efforts to save homes tend to become a costly diversion because they tie up firefighting forces and equipment that is needed to control the fire's growth. Firefighters can often save more homes if they avoid trying to save *every* home. Some fire departments are evaluating the WUI homesites in their jurisdiction to determine which ones they will protect and which ones they won't, due to poor access or other firefighting hazards.

Addressing a group of WUI homeowners, fire scientist Bob Mutch warned that the situation isn't "*if* a wildfire will burn your neighborhood"; it's "*when* will a wildfire burn your neighborhood?" One resident of a ponderosa pine forest verified Mutch's warning: "I grew up among the West Coast forests where fire danger was seldom rated very high, but after moving inland I was amazed to find that fire danger is very high or extreme for a large part of every summer!"

Wildfire risk is not like the risk from floods. Every year, there is one in a hundred chance of a hundred-year flood. The chance doesn't go up each year; it's still one in a hundred even if such a flood hasn't occurred for a century or longer. In contrast, following every year that a fire-dependent forest doesn't burn, the chance of a devastating wildfire increases slightly. Woody fuels accumulate continuously, year after year, unless measures are taken to reduce them. The only way to lessen the chance of a severe fire consuming your home and its surrounding forest is to regularly reduce hazardous fuels.

The importance of collective effort cannot be overemphasized. If you are the only one in a forest subdivision with a firewise homesite, after a wildfire your surviving home may sit within a devastated, dead forest. Steve Arno encountered a homeowner at his surviving summer residence after a fire in September 2000. The still smoldering wildfire had consumed most of the homes up the steep canyon and converted a dense forest of tall trees into a sea of charred skeletons, rising from black, ash-covered ground, now slimy due to recent rain. The only remaining green areas were small circles around two surviving homes, saved by the big agricultural sprinklers that sprayed

This house with a pine needle-covered shake roof would easily ignite if a wildfire burned through the surrounding forest.

water from irrigation ponds. Farther up the scorched canyon was a mobile home on a gravel pad, protected by an overarching steel roof. It survived but sat beneath 100-foot tall remnants of the incinerated forest. The owner, still in shock at the devastation, acknowledged a bit of good fortune that his vacation retreat was spared, but wondered if that was pointless since his lovely forest was now a disaster. He'd have to hire and oversee removal of a few hundred hazardous black snags and clean up the mountains of messy slash and debris. Then he would need to plant new trees and protect them from browsing by deer and elk. He knew the forest could eventually recover, but restoring good-sized trees would take so long that he questioned whether to keep the property. On the other hand, he reasoned, it would be hard to sell at any price in its desolate condition.

Collaborative efforts among private landowners and local governments are beginning to take shape around the West. The forested community of Sumpter in northeastern Oregon's Blue Mountains has made an impressive first step in ponderosa pine forest restoration. Harold Weaver, the early advocate of prescribed burning, was born and raised in Sumpter in the early 1900s when it was still a gold-mining town. Today the roughly 4 square miles that comprise the town's limits have only about 120 year-round residents but many summer homes and absentee landowners. Sumpter is surrounded by national forest, and in the late 1990s the Forest Service began thinning trees, piling slash, and underburning on its land bordering the town to reduce the threat of fire damage. The Oregon Department of Forestry, concerned about the dangerous, dense forest within the town limits, partnered with the county government and two rural fire departments. They sent multiple mailings to four hundred landowners, explaining the need for fuel reduction and cost-share opportunities, and provided informational meetings and a hazard assessment for each homesite. This effort mushroomed, and by 2010 about 1,000 acres among the small private properties had been thinned, and the slash either burned or chipped and hauled to a cogeneration plant.[6]

Most of the western states have programs to map the WUI and develop special fire-safe building requirements. Some counties have or are developing WUI zoning restrictions to prevent building in high-hazard areas, such as in narrow canyons or on steep slopes. Companies that offer homeowner insurance are using standardized criteria for evaluating a home's wildfire hazard. Some companies refuse or cancel coverage to homeowners in high-hazard situations, such as having a flammable roof or dangerous fuels surrounding the house. Real estate agents and appraisers are becoming aware of the importance of fire-resistant features for homes in the WUI. When a home is not eligible for fire protection or insurance, its value and the likelihood of qualifying for a mortgage are greatly reduced.

TEHAMA COUNTY

FIRE HAZARD SEVERITY ZONES
IN STATE RESPONSIBILITY AREA

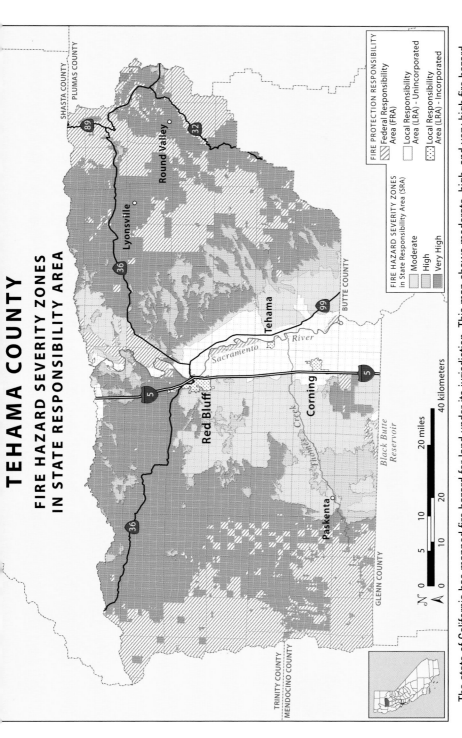

The state of California has mapped fire hazard for land under its jurisdiction. This map shows moderate, high, and very high fire hazard severity zones for state-protected lands in Tehama County, California. —*Map modified from California Department of Forestry and Fire Protection*

There is a slow trend of increasing involvement by county officials, homeowners, builders, and real estate and insurance representatives in creating safer homes and homesites in the WUI. This progress reflects broader recognition of the likelihood that a wildfire will threaten your forest home, but you can greatly increase the odds that it will survive. This realization should help prepare people living in the ponderosa pine forest to take action on the next step—protecting their home's surrounding forest in addition to their home itself.

By the 1980s rural fire departments and forest agencies throughout the West recognized they could not protect the proliferating numbers of homes springing up in fuel-rich forests, so they collaborated in efforts to alert residents to a harsh reality: "If you don't take measures to protect your home, we probably can't save it." Today, the western states provide websites, how-to literature, and consultations to help forest residents greatly enhance the fire resistance of their homes.[7] Also, the Firewise Communities website (www.firewise.org), maintained by the National Fire Protection Association, provides comprehensive information and recommendations for protecting homes from wildfire.

Creating a Firewise Homesite

Here is the scenario faced by thousands of WUI residents every year: The sheriff orders a final evacuation, and power to the area is shut off as a raging inferno driven by howling winds charges toward homes in the woods. The sickening roar of the advancing flames comes ever closer, and inky blackness replaces the daylight. An explosion punctuates the chaos—perhaps a homeowner's propane tank. Firefighters, stretched far beyond their limits, retreat from the towering flaming front. Your house and belongings will be history unless you already made them fire safe. How is that done?

1. The first step is to make your home's exterior fire resistant so it won't be ignited by firebrands lofted from the high-intensity flames torching trees and buildings hundreds of yards away. When the 1988 firestorm rampaged into the Old Faithful Resort in Yellowstone National Park, paved roads, expansive parking lots, and barren geyser formations kept flames well away from the historic Old Faithful Inn. But hundreds of burning embers rained down on the building's cedar shake roof. Valiant efforts of firefighters and equipment, including high-capacity, high-pressure hoses, continuously sprayed down the structure and saved it. The dozens of small rental cabins in the same complex couldn't be doused as frequently, and several of them were lost when their shake roofs burst into flame. Your home will not receive the VIP

protection that was lavished on the iconic Old Faithful Inn. Siding made of wooden shakes or rough-finished boards with cracks and knots invite ignition by firebrands carried as far as 1 mile from a crown fire. In contrast, asphalt shingles and plywood-based siding or even well-maintained log construction resists ignition from firebrands.

The home must also be completely separated from the forest's fuel bed, especially the continuous blanket of incendiary pine needle litter. Too often a house survives the blaze's flaming front but burns down later when firebrands ignite wood stacked against an outside wall or pine needles on the roof or under the deck. From August into November winds continually shower dry ponderosa pine needles onto decks, roofs, and gutters and up against buildings. They should be removed. Combustible landscaping like shredded bark, junipers, and dry grass must also be kept away from buildings. Just as residents of cities and suburbs maintain their yards, residents of the WUI must take care of their backyard, even if it is a forest.

The space around this fire-resistant house is clear of flammable shrubs, woodpiles, and pine needles.

2. The second step in home protection is to create a fuel-free ring around the house and other nearby buildings. On flat or gently sloping ground this should be at least 15 feet wide. If the house is on a steep slope the ring should be much wider below the house because fire-generated heat and wind flow upslope. This ring can consist of watered lawn and other noncombustible landscaping, gravel, asphalt, concrete, or stone, such as in a driveway, a parking pad, and walkways. Some deciduous shrubs and trees, such as aspen and maple, are basically nonflammable as long as their fallen leaves and dead stems are removed. Even the forest floor can be rendered relatively fire resistant by removing the litter, duff, dry grass, and other flammable ground cover like kinnikinnick, or bearberry.

3. The first two steps we've discussed minimize the wildfire's opportunity for direct contact with the house. A third step reduces the fire's intensity by limiting fuel as it approaches the home. Taming a crown fire and transforming it into a surface fire greatly reduces radiant heat and allows firefighters to safely stay and protect the home. A home should be surrounded with a low-hazard zone of forest about 120 feet wide. Trees within this zone should be thinned so their canopies are at least 10 feet apart. Branches within 10 feet of the ground should be pruned. Patches of small trees should be removed if located beneath or adjacent to overhead trees. Within this low-hazard zone, slash and flammable brush should be removed.

4. The fourth step for home protection provides safe access and exit in case of forest fire. Many rural and agency fire departments rate accessibility of woodland homes. Those deemed unsafe for fire engines and crews will not be defended. To qualify for protection, the homesite needs a turnaround and parking space adequate for a large fire truck and a few additional vehicles. The parking area needs generous overhead clearance from tree branches.

   The access road also needs to be safe. Trees should not overhang or encroach on a single-lane drive, clearance must be ample around curves, and pullouts should allow oncoming vehicles, including a fire engine, to pass each other. Also, the strip of forest bordering a narrow access road should be thinned out and cleared of slash. This treated corridor allows safer exit in case a blaze suddenly blows up or encircles firefighters, and it increases the road's potential to serve as a fire control line.

Woodland residents can make good use of common tools and techniques for protecting their homesite. A chainsaw with accompanying safety gear and knowledge of how to use it properly is a high priority for felling, bucking, limbing, and pruning trees. A pruning handsaw and pole pruning saw may be useful. To remove pine needle and grass litter, it is hard to beat the heavy plastic, clog-free leaf rakes available at some hardware stores and websites. A large aluminum grain scoop shovel functions as a long-handled dustpan for loading litter into a large garden cart, contractor-style wheelbarrow, pickup truck, or hauling trailer.

Whether or not homeowners burn their forest debris, it is important to have a few hundred feet of garden hose and a couple of good adjustable nozzles readily available. Small, accidental fires can be snuffed out promptly with a garden hose. A firefighter's 5-gallon backpack water pump is also very useful for quickly controlling a small fire that starts beyond reach of the garden hose.

Backpack water pumps are also basic tools for residents who choose to burn limbs and other forest debris in piles far away from the home's water system. In addition, they serve as a backup device in case a power outage (a common event) during wildfires prevents the pump in your well from working. People who have an all-terrain vehicle can equip it with an accessory water tank and pump for use in pile burning. Burning piles is an efficient way to dispose of slash and

Small trees located beneath big trees can act as ladder fuels, carrying surface fire into the crowns of larger trees. Pruning branches and cutting small trees can eliminate ladder fuels.
—*US Forest Service artwork*

litter, but it demands highly responsible behavior in terms of checking on regulations, obtaining a burning permit, consulting forecasts for wind events, notifying officials and possibly neighbors prior to ignition, having proper tools and a water source ready, tending the fire, and dousing its remains. Burners often get in trouble by leaving home under the assumption that their fires are completely out, only to have the wind fan them into a damaging wildfire for which they are liable.

It is safest to burn when the litter and the underlying duff are wet from rain or lightly covered with snow. Piles, or at least the middle of them, can be covered with a tarp to keep them dry. An inexpensive propane field-burning torch can get piles burning vigorously despite damp conditions. Pine branches and needles tend to burn readily because of their resin content. Piling for an efficient burn is an art. A core of dried-out limbs or wood chunks in the center helps, and green branches can be burned on top once a hot fire is achieved. Branches should be laid down in a compact pattern, mostly oriented in the same direction. Needle litter burns better if mixed with branches, allowing aeration. A pitchfork with five or six tines is handy for flipping wads of pine needles to keep them burning briskly, and for pushing the burning remains of branches together. You will also need a Pulaski, or a similar pick or shovel, to create a dirt fire line that encircles the burn pile and keeps flames from creeping out too far.

Another consideration for residential fire protection is developing a water supply that can be used to replenish a fire engine's tank or for the homeowner's own emergency irrigation sprinklers. The normal household water system won't provide water fast enough for refilling a pumper truck, and electricity is often cut off in advance of the fire. As the fire approaches, a pumper crew may wet down the house and immediate surroundings. If the pumper can access a water source, such as a swimming pool, pond, or water hole along a small creek, it can refill rapidly and provide more protection. Likewise, if homeowners have a gasoline-powered irrigation pump and a hose or small, farm-type irrigation system with sprinklers, they can create their own wet zone around the house. A water source as simple as a 1,000-gallon, above-ground poly tank can supply enough sprinkling water to keep fire farther away from buildings. Even a typical 400- to 500-gallon hot tub might be of some use.

One insurance company, Chubb, offers wildfire defense services as part of its homeowner policies marketed by independent agents in fourteen western states (see www.chubb.com). This insurance service provides a homesite inspection and advice for reducing wildfire hazard. If a fire threatens the area, the company dispatches firefighters in certified wildland engines to protect customers' homes. They remove fuel hazards adjacent to the house, set up sprinkler systems, and spray fire-suppressing gels on homes. Thirteen homes insured by Chubb were within the perimeter of the 2010 Fourmile Canyon fire in a ponderosa pine forest northwest of Boulder, Colorado. The fire destroyed 169 homes. Three of the homes insured by Chubb were destroyed, but the other ten were spared largely due to the extra protection provided by Chubb fire crews.

Safeguarding Your Forest

Once you've created a firewise homesite, it's time to think about safeguarding your forest. If a wildfire burns in the tree crowns, you'll be left with an ugly expanse of charred tree skeletons and ashy, baked soil. If you work to make the forest fire resistant, the ponderosa pines and the soil have a better chance of coming through a wildfire with little damage. If a fire burns only along the ground surface it is much less intense. Ground-cover plants will sprout from surviving root crowns, and erosion will be minimal.

Basic principles used for reducing flammability of trees surrounding the homesite can be modified and extended to make the entire woodlot more fire resistant. It isn't necessary to remove pine needle litter in the forest beyond the immediate homesite area. Thinning the trees to separate their canopies by 10 feet or more prevents a running crown fire. Removing slash and most of the understory trees that could serve as fuel ladders prevents torching of overhead canopies. Inhibiting canopy fire reduces fire intensity and the creation of airborne firebrands.

Woodland owners can get information, advice, and often an onsite visit from local foresters employed by their state's department of forestry or natural resources. A fire-hazard assessment helps landowners develop and compare alternative methods for making their property more fire resistant. Also, each state's USDA Cooperative State Research, Education, and Extension Service, headquartered at a state university, provides educational materials, workshops, and forest stewardship courses designed to educate landowners in forest ecology and options for protection and management.

Cost-sharing grants are available to help woodland owners finance thinning and slash disposal treatments that reduce the hazards of wildfire and damaging insects and disease. Foresters working for the extension service can furnish information about local contacts for cost-share programs. These programs reimburse forest owners for a portion of the cost of approved treatments and allow landowners to be reimbursed if they do the work themselves. Funding comes from the National Fire Plan through the US Department of Agriculture and Department of the Interior and from the Environmental Quality Incentives Program and Resource Conservation and Development Councils administered by USDA's Natural Resources Conservation Service.

Conservation easements are another potential source of financial assistance for people who own a sizable property and who will commit to keeping it as forest in perpetuity. Conservation easements are voluntary sales or donations of certain property rights. Typically the landowner either donates or sells the development rights to a conservation organization or a government entity that has a program to preserve undeveloped land, for instance as open space or wildlife habitat. Even donated easements may provide significant financial advantages to the landowner in terms of lowering income, property, and inheritance taxes. Conservation easements are entirely voluntary and are individually crafted to meet the needs of the original landowner and later owners of the property for managing the forest. The Land Trust Alliance (www.landtrustalliance.org) provides a clearinghouse of information about local organizations that grant conservation easements and offers publications that explain different aspects and alternatives involved with such easements.

It would be helpful if western states would follow the lead of several southern states that have established prescribed burning educational programs for forest landowners and adopted right-to-burn legislation, akin to the more widely known right-to-farm statutes.[8] Right-to-burn recognizes prescribed burning as a legitimate practice on forestland. However, cultural acceptance of forest burning is not established in the West like it is in the South, and because of prolonged exclusion of fire and the droughty environment, prescribed burning in the West is a bigger challenge than it is in the South.

By investing some time, energy, and money, residents of the ponderosa pine forest can transform their homesites into relatively safe places regardless of the ever-present threat of wildfires. They can also help safeguard their entire forest from damaging fire, insects, and disease. By thinning the trees and retaining the healthiest ones, they allow the remaining trees to grow big and have a long life. This intensive work produces a more natural ponderosa pine forest, akin to the giant pines and grassy glades admired by early explorers. Modern residents will admire the restored forests, too, and perhaps catch glimpses of pileated woodpeckers, black bears, or other forest inhabitants, all thanks to the stewardship of the forest landowner.

EPILOGUE

PONDEROSA, PEOPLE, AND FIRE—THE FUTURE

Americans always do the right thing—after they've tried everything else.

—Winston Churchill

THREATS TO THE WEST'S PONDEROSA PINE FORESTS are immediate, widespread, and severe. The problem is largely one of our own making, but it presents an historic opportunity to reverse a century of decline and restore more sustainable forests for the future. Previous activities—from clearfelling to high-grade logging to clearcutting to light thinning to fire suppression—are inconsistent with the long-term sustainability of ponderosa pine forests. Restoration treatments, in contrast, are based on ponderosa's ecology, specific attributes, and fire history. General approaches to restoration are known, and their effectiveness has been demonstrated. But the pace of restoration must quicken, and the scale of treatment needs to expand, goals that will require more than a few passionate individuals. This needs to be an all-hands-on-deck effort, including federal and state forestry organizations, residents of the wildland-urban interface, county agents, university extension and research personnel, volunteer fire departments, forestry consultants, homeowners associations, insurance companies, mortgage lenders, environmental organizations, and anyone else who values our western forests. Although many are already involved at some level, coordination is needed to attain critical mass and get the attention of decision-makers, including legislators. Smokey Bear's message needs to change from "Only *you* can prevent forest fires" to "Protect your home *and* Smokey's—restore the forest."

Over the next decade, extensive restoration treatments could rejuvenate millions of acres of highly threatened ponderosa pine forests. Developing strategically located zones of low-hazard ponderosa forests composed of increasingly larger trees will provide some breathing room to adapt more comprehensive restoration approaches. After a checkered past, we have the opportunity to work toward a positive future for the ponderosa forests that are so near and dear to us. All we need is the will to act.

PART II

PONDEROSA PINE ON AND OFF THE BEATEN PATH

Grand old Ponderosa you have set forth in magnificent style, describing its many forms . . . while showing their inseparable characters.

—John Muir, letter to Charles Sprague Sargent on June 7, 1898[1]

Observant travelers will find ponderosa growing in places ranging from bare rock to grassy savanna to moist forest, giving rise to trees of varying size, shape, and color and forests varying in density, pattern, and hue. Charles Sprague Sargent waxed eloquent about ponderosa—his favorite pine—in his 1897 *The Silva of North America*: "Possessed of a constitution which enables it to endure great variations of climate and to flourish on . . . well-watered slopes . . . on torrid lava beds, in the interior dry valleys of the north and on the sun-baked mesas of the south, and to push out over the plains boldly, where no other tree can exist." Part II of this book profiles some of these memorable ponderosa places in every western state and British Columbia, in landscapes as varied and colorful as Sargent described. We chose the sites based on their beauty, age, uniqueness, cultural significance, historical interest, or scientific value. The casual traveler, the inquisitive, and the seeker will all be richly rewarded for their efforts to visit these special places.

Featured ponderosa pine places in western North America.

Arizona

1. Rim Lakes Vista Trail—Apache-Sitgreaves NF
2. Rim Road—Coconino NF
3. Rincon Mountains East of Tucson
4. Chiricahua National Monument
5. Kaibab Plateau and Kaibab NF

California

6. Crystal Lake Recreation Area—Angeles NF
7. Eddy Arboretum at the Institute of Forest Genetics
8. Westwood
9. Big Bear Lake—San Bernardino NF
10. Pine Dunes Research Natural Area
11. Likely to Surprise Valley—Modoc NF

Colorado

12. Florissant Fossil Beds National Monument
13. HD Mountains Roadless Area—San Juan NF
14. McPhee Park—San Juan NF
15. Upper Beaver Meadows—Rocky Mountain NP

Idaho

16. Lake Coeur d'Alene Loop—Idaho Panhandle NF
17. Ponderosa State Park—Payette Lake
18. Ponderosa Pine Scenic Byway—Boise NF

Montana

19. Alta Ranger Station and Kramer Grove—Bitterroot NF
20. Big Pine—Lolo NF
21. Lick Creek—Bitterroot NF
22. Rimrock Drive—Custer NF
23. Missouri Breaks National Back Country Byway
24. Primm Meadow—Blackfoot Valley
25. Maiden Rock Area—Big Hole River

Nebraska

26. Bessey District—Nebraska NF

Nevada

27. Lee Canyon—Humboldt-Toiyabe NF
28. Hidden Forest—Desert National Wildlife Refuge
29. Wheeler Peak Scenic Drive—Great Basin NP

New Mexico

30. Mescalero Apache Indian Reservation
31. Santa Fe NF near Cuba
32. El Malpais National Monument
33. History Grove—Valles Caldera National Preserve
34. McKenna Park—Gila Wilderness
35. Monument Canyon Research Natural Area

North Dakota

36. Little Missouri National Grassland

Oklahoma

37. Black Mesa—The Panhandle

Oregon

38. Big Red—LaPine State Park
39. Heritage Forest Demonstration Area—Deschutes NF
40. Ukiah–Dale Forest State Scenic Corridor
41. Big Pine Campground—Rogue River–Siskiyou NF
42. Lost Forest Research Natural Area

South Dakota

43. Peter Norbeck Scenic Byway—Black Hills
44. Mt. Rushmore National Memorial

Texas

45. Davis Mountains Preserve
46. Guadalupe Mountains National Park

Utah

47. Albert Potter's Pines—Somewhere in Utah
48. Wah Wah Mountains
49. Red Canyon—Dixie NF
50. Toms Creek—Deep Creek Mountains
51. Hells Backbone—Dixie NF

Washington

52. Gifford Pinchot NF
53. White Pass Scenic Byway—Mt. Baker–Snoqualmie NF
54. Fort Lewis Area near Puget Sound
55. Turnbull National Wildlife Refuge
56. Desolation Peak—Ross Lake National Recreation Area
57. Leavenworth Loop—Wenatchee NF

Wyoming

58. Blacks Fork River—Mountain View
59. Vedauwoo—Medicine Bow NF
60. Mallo Camp—Black Hills
61. Pine Bluffs Rest Area

British Columbia

62. Hanging Tree—Lillooet
63. Anderson Lake Area—Squamish Forest District
64. Rocky Mountain Trench Ecosystem Restoration Program

ARIZONA

Rim Lakes Vista Trail—Apache-Sitgreaves National Forests

The Mogollon (pronounced *Muggy-own*) Rim, a 200-mile-long escarpment marking the southern edge of the Colorado Plateau in central Arizona, offers numerous vantage points overlooking the largest contiguous ponderosa pine forest in the country. Some of the best views are found along Rim Lakes Vista Trail #622, located about 30 miles east of Payson on AZ 260. Two popular, handicap-accessible overlooks are Rim Lake and Military Sinkhole. On a clear day, visitors are treated to panoramas of pine forests that resemble waves running to the horizon.

View southwest from the Rim Lakes Vista Trail on the Mogollon Rim, Apache-Sitgreaves National Forests.

Rim Road—Coconino National Forest

The Rim Road–General Crook Trail (Forest Road 300) provides impressive vistas of ponderosa forests in areas where the road skirts the edge of the Mogollon Rim. However, the best views are available from the end of short spur roads that head south on promontories off the main Rim Road. Part of the route follows the wagon trail that General Crook and his troops constructed in 1872. Martha Summerhayes, wife of an Army officer and author of *Vanished Arizona*, vividly described views from the rim while traveling the trail in 1874: "Surely I have never seen anything to compare with this—but oh! would any sane human being voluntarily go through with what I have endured on this journey, in order to look upon this wonderful scene?"[2] Visitors should be aware that Forest Road 300 (Rim Road) is rocky and washboarded in places. Visitors can access the Rim Road from the west by using AZ 260 from Camp Verde, Forest Road 3 from Flagstaff, or AZ 87 from Winslow. From the east, the Rim Road can be accessed by driving AZ 260 from Heber.

Rincon Mountains East of Tucson

The Southwest's expansive deserts are interrupted here and there by small mountain ranges called sky islands. One such island is the Rincon Mountains, which rise to an elevation of 8,664 feet in the Sonoran Desert just east of Tucson. Most of the Rincon Mountains is a wilderness area within Saguaro National Park. Calvin Farris, who conducted his PhD work in the Mica Mountain area of the Rincons, says forests here are likely the most frequently burned ponderosa forests in the country, with at least ten fires occurring since 1900.[3] Jim Malusa, a scientist at the University of Arizona, notes that fir are abundant on adjacent north slopes but absent on the relatively flat mountaintop—despite numerous draws moist enough to support them.[4] Fir's absence here is likely due to its lack of adaptation to frequent fires. Forests here also likely burned frequently before 1900. In late spring of 1886, Lt. John Bigelow and his troops combed the Rincons for nine days searching for Apaches when they received a dispatch from Fort Lowell, near Tucson: "Signal fires have been seen in the Rincon Mountains the last two nights." Bigelow responded, "These signal fires are doubtless the burning woods that have been observable to us ever since we came here."[5] The ubiquitous evidence of fire in the form of charred trees, logs, and open stands gives one a sense of being far back in time, and the grueling 13-mile hike up from the desert below adds a sense of being far away in space. Yet the night lights of Tucson belie the apparent remoteness of this wilderness. Farris still marvels at how this small landscape near Tucson became a relict for relatively pristine, burned ponderosa pine stands. The Mica Mountain area can be reached via the Douglas Springs Trail, which takes off at the east end of Speedway Boulevard in southeast Tucson. Another, slightly longer route is the Tanque Verde Trail, which departs from the Cactus Forest Loop Drive about 1 mile southeast of the Saguaro National Park visitor center. The center is located at 3693 South Old Spanish Trail in Tucson (520-733-5100). Permits are required for overnight camping in the park.

Chiricahua National Monument

An obscure sky island in the southeast corner of Arizona is home to Chiricahua National Monument. The monument's striking rhyolite rock formations, called hoodoos, have earned it the nickname Wonderland of Rocks. The Apaches called it the Land of Standing-Up Rocks, and the rugged terrain served as a stronghold for Cochise and Geronimo. Typically, ponderosas in dry environments grow in nearly pure stands, as amply demonstrated across much of northern Arizona. Yet here, in the driest conditions in which ponderosa can grow, it typically occurs in a mixture with other pines. Three of them are pinyons—pinyon pine, border pinyon, and singleleaf pinyon—and two others are pines common to Mexico, Apache pine and Chihuahuan pine. Arizona pine, whose occurrence is restricted to about a dozen mountain ranges in the Southwest, is also found here. This typically four- or five-needled pine was historically viewed as a variety of ponderosa pine (var. *arizonica*) but is now recognized as a separate species. The 2011 Horseshoe 2 wildfire severely impacted the southeast portion of the monument, but ponderosas can still be found among the hoodoos along the upper part of Echo, Rhyolite, and Jesse James Canyons. The monument can be reached by driving south from Willcox for 33 miles on AZ 186, and then east for 4 miles on AZ 181.

Ponderosas growing among the hoodoos in Chiricahua National Monument. The gray skeletons in the middle of the picture are pines killed by the 2011 Horseshoe 2 wildfire.

Kaibab Plateau and Kaibab National Forest

Perhaps no place in the West is better known for its ponderosa pine forests than the area surrounding the Grand Canyon. Traveling northwest from Flagstaff to Grand Canyon Village (South Rim), and from Jacob Lake south across the Kaibab Plateau via AZ 67 to the canyon's North Rim, showcases an impressive variety of ponderosa pine trees and forests. John Faris, author of the 1930 book *Roaming the Rockies*, recalled an encounter he had with an early day traveler and camper in the Grand Canyon area. "I've liked your Canyon, and I've liked your deer," the man told Faris, "but, best of all, I think, is your Kibosh [sic] Forest."

CALIFORNIA

Crystal Lake Recreation Area—Angeles National Forest

Blue-green Crystal Lake, pine forests, campgrounds, and hiking trails can be found in a high mountain basin just out the backdoor of millions of Los Angeles residents. This pine hideaway can be reached by driving 25 twisting miles up the San Gabriel Canyon east from Azusa on CA 39. The only naturally occurring lake in the San Gabriel Mountains, Crystal Lake got its name in 1887 when a Pasadena judge claimed, "The water is clear as a crystal, and our party found it good to drink."[6] More recently, the once-emerald waters have been degraded and diminished by years of drought—compounded by ash- and debris-laden runoff from the 2002 Curve

Pines along the road in Crystal Lake Recreation Area.

wildfire. The lake's waters are gradually regaining clarity due to inflow from underground springs and installation of an aeration system. Most of the surrounding pine forest remains intact, however, and continues to impress visitors. Both ponderosa pine and Jeffrey pine occur here, but the observer will be hard-pressed to tell the difference without picking up the cones. Ponderosa cones are typically prickly, whereas Jeffrey cones are not.

Eddy Arboretum—Institute of Forest Genetics in Placerville

In 1925, after consulting with renowned fruit geneticist Luther Burbank, lumberman James Eddy established the first facility in the world exclusively focused on forest genetics and tree breeding. Because he was running short on money, Eddy donated the Institute of Forest Genetics to the federal government in 1935. The arboretum features ponderosa pines planted from seeds collected from fifty different geographical areas across western North America.[7] Research at the institute focuses on testing elevational races of ponderosa pine to determine appropriate matching of seeds and sites for reforestation, information that will be critical in a changing climate. It also claims the best collection of pines from around the world. A self-guided trail through the arboretum makes the wide range of plantings accessible to visitors. Those with in-depth interest in ponderosa pine genetics can call ahead and arrange a guided tour. The arboretum is located at 2480 Carson Road (at the top of the hill) about 2 miles east of Placerville (530-622-1225).

Westwood

After acquiring 900,000 acres of prime ponderosa pine–mixed conifer timberlands in northeastern California, Minnesota timber baron T. B. Walker started building the company town of Westwood in 1912. Westwood was a dry town and Walker's Red River Lumber Company controlled virtually all aspects of community life there. Red River Lumber soon hired a man to develop an advertising campaign based on the Paul Bunyan character, and in 1914 published an advertising pamphlet, "Introducing Mr. Paul Bunyan of Westwood, California."[8] The town's colorful history is depicted in its forestry museum, alongside a giant statue of Paul Bunyan and Babe the Blue

Ox. Walker was a pioneer in practicing light burning of the forest understory to control fuel accumulations in the company's pine forests, and he invested much time and money attempting to get the Forest Service to do the same. Westwood is located in the heart of pine country, midway between Chester and Susanville along CA 36.

Paul Bunyan and Babe the Blue Ox, outside the forestry museum in Westwood.

Big Bear Lake—San Bernardino National Forest

Before Euro-American settlement in the 1800s, the country around Big Bear Lake was teeming with bears and home to the Serrano Indians. The Serranos used the term *Yuhaviat*, which translates into Pine Place, to describe the area.[9] Both ponderosa pines and look-alike Jeffrey pines are found here. Big Bear Lake can be reached by driving 45 miles northeast from San Bernardino via CA 18, including a white-knuckle section through Rimforest and Skyforest that provides huge views to the west.

Pine Dunes Research Natural Area

Along the northeastern edge of California, an isolated stand of fewer than one hundred ponderosa pines grows in loose sands that originated from an ancient lakebed. There are several old trees (up to 57 inches in diameter) in the stand and clumps of younger trees, but there is no evidence of reproduction within the last sixty years. The BLM designated the stand a Research Natural Area in order to generate interest among scientists to study the stand's origin and uncertain future.[10] Golden eagles have nested in a dead-topped pine at the western edge of the stand, and alert visitors may see mule deer, pronghorns, and wild horses. The trees can be reached by traveling about 17 miles east of Ravendale via the BLM's Buckhorn Backcountry Byway, Marr Road, and a short spur road. Visitors are strongly encouraged to contact the BLM Eagle Lake Field Office at 2550 Riverside Drive in Susanville (530-257-0456) for detailed directions to this out-of-the-way relict stand (T35N, R16E, southern portion of Section 25), part of which falls on privately owned land.

Likely to Surprise Valley—Modoc National Forest

A drive from the town of Likely on the west side of the South Warner Mountains to Surprise Valley on the east (via Forest Road 64 and County Road 258) traverses miles of forests with ever-changing species composition. Impressive large, old pines are scattered along the way, and ponderosas can be seen growing on a wide range of sites and with numerous associates, including juniper, white fir, aspen, and even lodgepole pine at the higher elevations. There are some

Large ponderosas growing within an aspen clone along the road from Likely to Surprise Valley. Shade-tolerant firs and spruce, rather than ponderosa pine, are more commonly seen growing with and gradually displacing shade-intolerant aspen.

magnificent Washoe pines at the Patterson Campground, which is a short half-mile detour northeast from the main road. Washoe pine is very similar to ponderosa but has slightly smaller cones that are not as prickly. Its lineage is controversial—some view it as a separate species, while others consider it a variety of ponderosa pine (var. *washoensis*). Those starting from the Surprise Valley (east) side should watch for Forest Road 64 and County Road 42, which head south and then west from County Road 1 (Surprise Valley Road) near the north end of Lower Alkali Lake.

COLORADO

Florissant Fossil Beds National Monument

A small number of old-growth ponderosas growing at Florissant Fossil Beds National Monument have special cultural significance. These pines bear unusual scars made long ago when the Ute Indians used them as medicine trees. A single 6- to 12-inch horizontal cut was made on the tree at a height that correlated with the location of the illness on the patient's body. A sharp stick was then inserted into the cut and lifted upward to peel the bark away, and the exposed inner bark was removed and used in a healing ceremony.[11] These scars differ from the far more common and larger scars that result from peeling bark for food. The monument also features the

A medicine tree at Florissant Fossil Beds National Monument.

Ponderosa pines growing around the fossilized stump of an ancient redwood at Florissant Fossil Beds National Monument.

bizarre juxtaposition of ponderosa pine growing in a dry, cold basin at 9,000 feet elevation among giant petrified stumps of redwoods that grew here over 30 million years ago when it was moist and warm. Florissant Fossil Beds National Monument is located 30 miles west of Colorado Springs via US 24, then about 2 miles south on Teller County Road 1.

HD Mountains Roadless Area—San Juan National Forest

The old-growth ponderosa pine stands tucked away in this remote part of the San Juan National Forest in southwestern Colorado will reward those willing to get away from the crowds. Rugged canyons are a distinguishing feature of the HD Mountains, named after an 1800s cattle brand, and old-growth pine stands found in Ignacio Canyon are especially impressive—some say the most impressive in Colorado. Laurie Swisher, old-growth specialist with the San Juan National Forest, says her favorite stand in the HDs spreads across a broad mesa.[12] Because of the absence of grazing, the stand features a native grass understory rather than a Gambel oak understory, and it shows no evidence of past logging, which is uncommon for stands on relatively flat terrain. The HD Mountains Roadless Area is located southeast of Bayfield. Directions to the old-growth stands in the HDs can be obtained by contacting the San Juan National Forest at 15 Burnett Court in Durango (970-247-4874). Interested visitors should note that some roads may be closed due to energy development, and that access may require permission to cross tribal or private lands.

McPhee Park—San Juan National Forest

McPhee Park was set aside in 1925—a farsighted move at the time—based on an agreement between the New Mexico Lumber Company and the Forest Service to preserve a sample area of virgin ponderosa pine timber.[13] A trail makes a loop through the park's old trees, but it may be difficult to find due to recent management activities. Visitors should be aware that some suppressed younger trees as well as older dead trees are occasionally removed from the area to reduce hazard and provide more growing space for the remaining trees. McPhee Park can be reached by turning north off Central Avenue in Dolores onto Eleventh Street (also called Dolores-Norwood Road, Road 31, and Forest Road 526), then driving about 14 miles to McPhee Park (left on Forest Road 531). Although there is an informational sign within McPhee Park, the turnoff into the area is not well signed, so visitors are encouraged to stop at the Dolores Ranger Station for detailed directions.

Upper Beaver Meadows—Rocky Mountain National Park

Some of the largest ponderosa pines along Colorado's Front Range can be found in the Upper Beaver Meadows area of Rocky Mountain National Park. Peter Brown, director of Rocky Mountain Tree-Ring Research, cites it among his favorite old-growth places.[14] The area features large old pines in a savanna setting, interspersed with meadows and occasional aspen groves. Longs Peak, elevation 14,259 feet, looms to the south. The Upper Beaver Meadows area is located about 5 miles west of Estes Park via CO 36 and Upper Beaver Meadows Road, which also traverse appealing pine savannas.

View looking over Upper Beaver Meadows toward Longs Peak in Rocky Mountain National Park.

IDAHO

Lake Coeur d'Alene Loop—Panhandle National Forest

Located in Idaho's Panhandle, Lake Coeur d'Alene provides a sparkling contrast to the soft, deep-green foliage of the ponderosa pines that surround it. Part of the drive around the lake includes the Coeur d'Alene Scenic Byway, which at times rises high above the lake's shimmering waters. For the hiker, the North Idaho Centennial Trail extends for several miles east along the lake and west along the Spokane River from the waterfront in downtown Coeur d'Alene. There are additional trails in Heyburn State Park near the south end of the lake. Bikers can ride the Trail of the Coeur d'Alene, a paved bike path that skirts the southern part of the lake.

Ponderosa State Park—Payette Lake

Ponderosa State Park is located on a large peninsula jutting into Payette Lake just a few miles northeast of McCall. The scenic drive to the Narrows Overlook at the tip of the peninsula is a must. Visitors will

View from the Narrows Overlook at Ponderosa State Park near McCall.

see plant communities ranging from sagebrush to moist fir forest, but it is the magnificent ponderosa pines, some with colorful trunks up to 5 feet thick, that are the drawing card here. Historically, periodic surface fires favored fire-tolerant ponderosa pine over Douglas-fir and grand fir. However, a century of fire exclusion and the moist growing conditions in the park have allowed shade-tolerant firs to establish in the understory and compete with the pines. Park managers have responded by conducting thinning and prescribed burning treatments in parts of the park to open the pine stands and remove most of the firs. Similar treatments will be needed periodically and over larger areas to ensure that Ponderosa State Park continues to live up to its name.

Ponderosa Pine Scenic Byway—Boise National Forest

The Ponderosa Pine Scenic Byway is a 130-mile section of ID 21 running northeast from Boise to the town of Stanley. The prime section

South Fork of the Payette River along the Ponderosa Pine Scenic Byway.

of the byway for viewing ponderosa pines starts a few miles north of Lucky Peak Reservoir and continues to about 20 miles east of Lowman, where the ponderosa pine forest gradually gives way to lodgepole pine and fir with increasing elevation. East from Lowman, the turquoise waters of the South Fork of the Payette River complement the deeper green hues of the pines along its banks. Near Idaho City, travelers will also see ponderosas growing on piles of soil and rock that were left after dredge mining about a century ago.[15] Other highlights are the bizarrely shaped, small ponderosas and gnarled root systems exposed at several roadcuts along the way, particularly where the road crosses high ridges.

MONTANA

Alta Ranger Station and Kramer Memorial Pine Grove—Bitterroot National Forest

The Alta Ranger Station, embedded in a beautiful ponderosa forest, is the oldest Forest Service ranger station in the country, dedicated on July 4, 1899.[16] Two rangers constructed the sod-roofed, 13-foot by 15-foot log building in just two weeks and bought the supplies with their own money. Some of the most impressive old ponderosas at Alta can be seen in the nearby Kramer Memorial Pine Grove. The Alta area also features scattered old pines with Indian bark-peeling scars. The Alta Ranger Station and Kramer Grove are 35 miles southwest of Darby via US 93 and MT 473 (West Fork Road). Another concentration of bark-peeled pines can be seen at Indian Trees Campground, 4 miles south of Sula and 1 mile west of US 93.

Big Pine—Lolo National Forest

The largest ponderosa pine known in Montana—called Big Pine—is located west of Missoula along the banks of Fish Creek. Big Pine is nearly 7 feet in diameter and 194 feet tall. The big tree can be reached by driving about 40 miles west from Missoula on I-90 to exit 66 (Fish Creek Road), and then south about 4 miles to the Big Pine Fishing Access Site.

Lick Creek—Bitterroot National Forest

A self-guided auto tour of the Lick Creek Research/Demonstration Area in the Bitterroot Valley features a variety of actively managed ponderosa pine stands that demonstrate thinning, prescribed burning, and uneven-aged (selection) management. The Lick Creek drainage was the site of the first ponderosa pine timber sale on national forest lands in Montana (1906–1909), a sale that was visited and reviewed by Gifford Pinchot. Lick Creek is also home to a series of photo points that have been periodically photographed since 1909. A tour brochure and directions to the Lick Creek–Lost Horse loop road (beginning and ending on US 93) can be obtained from the Bitterroot National Forest offices in Hamilton (406-363-7100) or Darby (406-821-3913).

Rimrock Drive—Custer National Forest

Ponderosa pine forests are widely scattered across the broken country of southeastern Montana. One notable area of pine forest can be found near the small town of Ekalaka. A good access point is the Stagville Draw Road (Forest Road 3813), which heads west and then south from MT 323 about 3 miles south of Ekalaka. About 5 miles in, Rimrock Carter Road (Forest Road 3104) takes off to the east and follows a ridgeline for several miles, providing handsome views of ponderosa forests, rimrocks, and the Chalk Buttes to the southwest. As the road descends and turns back to the north toward MT 323, it passes Camp Needmore, a Civilian Conservation Corps camp that operated from 1935 to 1946 as crews built infrastructure in the area. It is still used today for family reunions, 4-H camps, and celebrations.[17]

Missouri Breaks National Back Country Byway

Montana's Missouri River Breaks are a unique geological phenomenon featuring sandstone cliffs, eroded coulees, and ponderosa pines. But these are not ordinary ponderosas. Pines growing here are typically short and stout, with old trees taking on a broccoli look that reflects a land of scant precipitation, strong winds, and widely fluctuating winter temperatures. In 1883, historian Eugene Smalley referred to such trees as "little dwarf pine in ravines."[18] These pines grow at the crossroads of history. Part of the byway (Two Calf Road north from Knox Ridge Road) parallels the Nez Perce National Historic Trail. Turning east, it also passes above campsites used by Lewis and Clark on their journey to the Pacific. Driving the byway is best done in a four-wheel-drive vehicle and should not be attempted in wet weather, when any vehicle can bog down in sticky clay gumbo. Those traveling east on the byway will start from MT 236 at Winifred, while those heading west will begin the byway from MT 191 near James Kipp Recreation Area on the Missouri River. Visitors are encouraged to check in at the BLM Lewistown Field Office, 920 NE Main Street (406-538-1900) for detailed directions and up-to-date road conditions.

Scattered ponderosas along the Missouri Breaks National Back Country Byway. Cliffs of the upper Missouri River loom in the background. —Photo courtesy Bureau of Land Management

Primm Meadow—Blackfoot Valley

The old-growth ponderosa pine stand at Primm Meadow northeast of Missoula has a long history with people. Salish tribal elder Louis Adams speaks about his tribe traveling across the mountains in spring to peel bark from the large ponderosas and occasionally burn the area. The original homesteader, and later the Primms—who bought the property in 1938—didn't cut the trees but did burn and mow under them to stimulate grass for their livestock. The land eventually passed to Plum Creek Timber Company, which signed a permanent conservation easement on the 112-acre tract in 2005.[19] A 2003 wildfire severely damaged the surrounding dense second-growth forests, but flames dropped to the ground and burned through the open-grown Primm Meadow pines, and the picturesque old trees survived. Directions to the trailhead for the nearly level, 2-mile hike to Primm Meadow are available at the Five Valleys Land Trust office in Missoula. A video documentary, "Pines of Primm Meadow," is also available from the Five Valleys office.

The old pines of Primm Meadow along the West Fork of Gold Creek in the Blackfoot Valley. —Photo by Kristi DuBois

Maiden Rock Area—Big Hole River

A geographically isolated population of ponderosa pines can be found in southwestern Montana, well outside the previously known range for this species. This relict population occurs as scattered individuals or small clumps strung out along the Big Hole River, beginning about 0.6 mile downstream from the Maiden Rock bridge. The population is uneven-aged, with a few trees over 630 years old. BLM-sponsored research shows that trees in the stand are highly inbred, which is not unusual for an isolated population. What is unusual is that the trees share the same genetic lineage as ponderosas in northern California, eastern Oregon, and central Idaho, but differ from those in both eastern and western Montana.[20] Emily Guiberson, a BLM forester who has collected samples from the stand, cautions that most of the area is privately owned, including the railroad bed.[21] She recommends floating the river as the best option for seeing these out-of-the-way trees. The Maiden Rock bridge can be reached by taking the Moose Creek exit (exit 99) from I-15 about 4 miles south of Divide, then driving west about 2 miles to the river.

NEBRASKA

Bessey District—Nebraska National Forest

Beginning in 1890, Charles Bessey, a professor at the University of Nebraska, urged the federal government to plant trees in the Sand Hills, a treeless region of north-central Nebraska. Settlers had moved into the area following the Homestead Act of 1862 to claim their 160 acres but found the soils too sandy to raise crops. The free acreage was later upped to 640, but old-timer Caroline Sandoz Pifer remained skeptical: "A person couldn't even starve properly on 160 acres, but on four times the land could starve just fine."[22] Though the land was unsuitable for farming, Bessey believed that planting trees would help control erosion, provide a source of fuel and fence posts, and create a sanctuary for wildlife. After repeated coaxing, the Federal Division of Forestry established a small ponderosa pine plantation on a local ranch in 1891. Bessey continued lobbying and finally convinced President Teddy Roosevelt to create the Dismal River Forest Reserve in 1902.[23] The reserve was later renamed the Bessey Ranger District of the Nebraska National Forest. Approximately 60,000 acres on the Bessey District have been planted since 1902—primarily with ponderosa pine—transforming what was once mixed-grass prairie into a national forest.[24] This effort has been variously described as "constructing a technological forest" and as an early-day example of "environmental engineering."[25] Bessey Nursery, located 1 mile west of Halsey on NE 2, is the oldest federal tree nursery in the United States. It has provided most of the trees for planting the Bessey Ranger District and made millions of additional seedlings available to other national forests in Colorado, Nebraska, South Dakota, and Wyoming. The best vantage point for viewing the ponderosa forest is the observation deck of the Scott Fire Lookout Tower, located about 2 miles southwest of the nursery.

View of planted ponderosa pine forest looking west from the deck of the Scott Fire Lookout Tower near Halsey, Nebraska.

NEVADA

Lee Canyon—Humboldt-Toiyabe National Forest

The road up Lee Canyon (NV 156) heads west from US 95 just 25 miles northwest of Las Vegas. Scattered ponderosas are first encountered as the road climbs above 7,000 feet elevation near the

Ponderosa pine in Lee Canyon near the ski area at Mt. Charleston.

junction with NV 158 (Deer Creek Highway). Numerous large pines soon appear and provide an impressive colonnade for the remainder of the drive, which terminates at the Las Vegas Ski and Snowboard Resort at the base of Mt. Charleston.

Hidden Forest—Desert National Wildlife Refuge

The 1.5-million-acre Desert National Wildlife Refuge is the largest refuge in the Lower 48 states. Despite few water sources, the primary
north-south backbone of the refuge—the Sheep Range—supports
some remarkable old-growth pines. Ponderosa pine adaptations to

*Five-foot diameter ponderosa with a massive fire scar, growing
among other old-growth pines in Hidden Forest. Debris from flash
flooding has dislodged patches of bark at the base of the tree.*

the desert environment include shorter needles and small cones—some much smaller than pinyon cones.[26] Ponderosas can be found up to nearly the highest elevation (9,990 feet) of the Sheep Range, where they mingle with bristlecone pines. Of special interest here is Hidden Forest, an impressive stand of old-growth pines spread along the floor of Deadman Canyon. The trailhead to Hidden Forest can be reached by driving about 20 miles north from Las Vegas on US 95, east 4 miles on the Desert National Wildlife Refuge road (Corn Creek Road), north 14 miles on Alamo Road, then east about 4 miles on Hidden Forest Road to the trailhead. Cooler spring or fall days are recommended for this grueling trip, and visitors should be forewarned that access roads are rough and washboarded, with sections of loose rock. The Hidden Forest is a gradual uphill hike from the trailhead, with the first ponderosas encountered at about the 3.7-mile mark. Those who make this challenging hike will be rewarded by one of the most appealing and interesting ponderosa pine stands anywhere. The incongruity of finding this ponderosa pine oasis just a few miles from the desert further adds to its allure. Hidden Forest is dominated by old-growth trees, many of surprising size, but healthy pines of all sizes are present. Fire scars are common, even though most trees grow on exposed gravel on the canyon floor or on rock along the sides. An old warden's cabin can be found about 5.7 miles from the trailhead, complete with a small spring nearby.

Wheeler Peak Scenic Drive—Great Basin National Park

The best way to see ponderosa pine in Great Basin National Park is to drive the entire 10-mile length of the Wheeler Peak Scenic Drive. However, do not come to this remote national park expecting to see extensive ponderosa pine stands or large, eye-catching, old-growth trees. Instead, appreciate the isolated population of pines that manages to perpetuate itself here at the margins of the species' range. The first individual ponderosas along the scenic drive are encountered at about 6,800 feet, and then pines begin appearing in small clumps at Upper Lehman Campground. With increasing elevation, the scenic drive bisects intriguing multicolored, multitextured forests, where ponderosas grow amidst Douglas-fir, white fir, Engelmann spruce, mountain mahogany, singleleaf pinyon pine, and aspen. Perhaps the best place in the park to see ponderosas resembling a stand is below the road at Osceola Ditch. Pines in modest numbers can also be seen on the north side of the road at Mather Overlook (about 9,000 feet elevation), but ponderosas reach their cold limits soon thereafter at about 9,300 feet. The last leg of the scenic drive turns south to the road's terminus at about 10,000 feet, providing a full-on view of the massive north face of 13,064-foot Wheeler Peak. Another attraction is the Bristlecone Pine Trail, a 3-mile hike that takes off from the terminal parking lot and features ancient, weather-sculpted bristlecone trees thousands of years old. Wheeler Peak Scenic Drive can be accessed by driving about 4 miles west from Baker on NV 488 (Lehman Caves Road).

NEW MEXICO

Mescalero Apache Indian Reservation

In 1914, there was a movement to turn the entire Mescalero Apache Indian Reservation in southeastern New Mexico into a national park.[27] The park did not come to fruition, but the reservation today is a heavily used recreation area that includes sightseeing, hunting, downhill skiing at Ski Apache, and a destination resort-casino called Inn of the Mountain Gods. Approximately 150,000 acres of the reservation are also considered commercial forest—much of it ponderosa pine. The forests are managed on a twenty-year cutting cycle using uneven-aged (selection cutting) methods, with about 7,500 acres harvested annually. These intensively managed ponderosa forests are both productive and visually appealing. The reservation is located south and west of Ruidoso along US 70 and NM 244.

Santa Fe National Forest near Cuba

An isolated stand of purple-barked ponderosa pines straddles both sides of US 550 a few miles northwest of Cuba. These trees, along with a few scattered pines just south of town, are remarkable in that they are the only forest trees along the 400-mile drive on I-25/NM 550 from Las Cruces in southern New Mexico to Aztec in the north.

The sparse understory of this stand consists of a mosaic of smooth, light yellow to rust-colored rock, pine needles, and occasional clumps of grass. Picturesque, large old ponderosas are sprinkled below and among the small sandstone cliffs that define the area. It is unclear whether the pastel purple bark is due to the chemical composition of the substrate on which the trees grow, the genetic makeup of this isolated population, or some other factor. A good place to walk among these pines is about 0.2 mile northwest of the junction of NM 550 and NM 96. Visitors will need to climb through a barbed wire, right-of-way fence on either side of the highway to see the stand close-up.

The ponderosa pines growing among small sandstone cliffs near Cuba have light purplish bark.

El Malpais National Monument

The word *malpais,* meaning "badlands" or "bad country," was used by Spanish explorers to describe lava rock terrain. The young volcanic lands at El Malpais National Monument have little soil, and the trees that grow on the lava flows are stunted and malformed. The Zuni-Acoma Trail traverses large areas of lava terrain and is a good route for viewing the dwarf, bonsai-shaped pines. The Big Tubes and Lava Falls areas are other good places to see miniature, oddly shaped trees. Patterns of tree-ring widths collected from El Malpais's ancient trees—some of the oldest in the Southwest—have been used by scientists to help establish the relationship between tree-ring patterns and climate.[28] A remnant of a ponderosa pine tree found on the monument's Bandera lava flow was dated back to the year AD 111. El Malpais National Monument can be reached by driving south from US 40 near Grants on either NM 53 or NM 117.

Stunted, bonsai-shaped ponderosas growing in cracks of lava rock at El Malpais National Monument. These trees struggle to survive because of highly restricted root systems.

These large ponderosas grow near the dwarf pines at El Malpais but just off the lava flow, illustrating the importance of well-developed root systems to growth.

History Grove—Valles Caldera National Preserve

The History Grove, an extensive old-growth ponderosa stand that runs along the base of Redondo Peak in the Valles Caldera National Preserve, was either missed or deliberately avoided during early-day logging. A few of the old pines have crosses carved into them, suggesting that the area was used as a place of worship by early sheepherders.[29] The carvings may have protected this impressive stand from later cutting, as much of the surrounding forest was heavily logged. Some old trees in the grove also exhibit Indian bark-peeling scars. The History Grove can be reached by driving about 20 miles west from Los Alamos via NM 501 and NM 4, and then a few miles north to the preserve headquarters. Travelers should call ahead for visitation hours.

McKenna Park—Gila Wilderness

For the fit and adventuresome, extensive old-growth ponderosa forests at McKenna Park in southwestern New Mexico provide a unique hiking or horseback-riding destination. The stands lie in the Gila Wilderness, which was dedicated in 1924, making it the oldest wilderness area in the United States. McKenna Park old-growth stands continue to be shaped by fire, as the Gila was one of the first wilderness areas in the country to adopt the policy of letting some natural fires burn. Detailed guidelines for implementing the new policy were written by a twenty-two-year-old firefighter named Lawrence Garcia in the early 1970s, based on his familiarity with the wilderness and experience fighting fire.[30] Repeated burning has reduced fuels and made much of the area more resistant to wildfire. Directions to the McKenna Park old-growth stands can be obtained by contacting the Gila National Forest at 3005 E. Camino del Bosque in Silver City (575-388-8201).

Monument Canyon Research Natural Area

A 650-acre, old-growth ponderosa pine stand in Monument Canyon in the Jemez Mountains was established as a research natural area in 1932. Elders of the Jemez Pueblo knew the canyon as the Place Where the Clouds Live. Historically, surface fires occurred in the stand on an average of every six years for at least 300 years until 1892, the year of the last recorded widespread fire.[31] Lack of fire allowed a dense sapling understory to develop, putting the entire stand at risk. To reduce the threat, about half of the research natural area has received thinning and burning treatments since

2006. Monument Canyon Research Natural Area is part of the larger 7,000-acre San Juan fire management area, which receives regular prescribed burning to maintain open stand conditions. The size, age, and cultural significance of the ponderosa stand at Monument Canyon rank it among the most significant in the Southwest.[32] Visitors should contact the Jemez Ranger Station in Jemez Springs (575-829-3535) for current road conditions and alternative routes to this remote site. One scenic but circuitous route, NM 290 to Forest Road 10, passes through the small town of Ponderosa.

NORTH DAKOTA

Little Missouri National Grassland

The farthest northeastern extension of naturally occurring ponderosa pines can be found in scattered stands along the Little Missouri River in southwestern North Dakota. The trees in this harsh

Ponderosas overlooking the Little Missouri River northwest of Amidon.

environment are restricted to coarse-textured soils, exposed shale, or limestone outcrops. The largest and oldest ponderosas documented in the area are about 70 feet tall, 2 feet in diameter, and nearly 350 years old. The uniqueness of these outlier stands spurred establishment of the 13,940-acre Dakota National Forest in 1908, although the designation was abolished a few years later.[33] The Little Missouri ponderosa stands can be accessed by driving the East River Road, which takes off from ND 85 about 2 miles west of Amidon. This route first heads west for about 1 mile, then turns north and continues for about 10 miles to a fork. The left fork, which heads southwest, provides the most extensive views of ponderosa pines.

OKLAHOMA

Black Mesa—The Panhandle

Ponderosa pines can only be found in one locale in the Sooner State—near Black Mesa in the Oklahoma Panhandle, about 25 miles west of Boise City.[34] Black Mesa rises about 600 feet above the surrounding plains and boasts the state's highest elevation at 4,973 feet. Ponderosa's occurrence is limited to a small piece of privately owned land in upper South Tessequite Canyon, just a few miles west of Black Mesa State Park.[35] Owners of the canyon are currently not letting anyone on their property to see the trees, however.

OREGON

Big Red—LaPine State Park

LaPine State Park is located about 3 miles west of US 97 via State Recreation Road, just north of La Pine. It is home to the largest ponderosa pine in the state, based on a standardized scoring formula that includes height, diameter, and crown spread.[36] Known as Big Red, the champion tree measures 167 feet in height and 9 feet 3 inches in diameter (which makes it the largest-diameter ponderosa in the country), and has a crown spread of 68 feet. Managers believe the bumpy-stemmed, rugged-looking veteran was spared the axe in early-day logging because of weather damage that reduced its value for lumber. Thinning treatments are ongoing in various parts of the park and adjacent ponderosa pine forests to help protect Big Red and the surrounding area from severe wildfire.

Heritage Forest Demonstration Area—Deschutes National Forest

The 14-mile drive from Sisters to the Heritage Forest Demonstration area provides several ponderosa pine highlights. The first leg, a 9-mile segment of US 20 west from Sisters to the Camp Sherman Road, threads its way through beautiful ponderosa forests. An old-growth stand opposite the Black Butte Resort is especially impressive. The primary feature, however, is the Heritage Forest Demonstration area about 5 miles north of US 20 on Camp Sherman Road. The demonstration plots begin at the Camp Sherman Resort entrance and extend for about 1 mile north to the Allingham junction (Forest Road 1217). This cooperative project between Friends of the Metolius and the Deschutes National Forest includes informational signs that describe the various thinning and burning treatments being tested for restoring the area's ponderosa forests. Details of the project are available at http://www.metoliusfriends .org/activities.html.

Ukiah–Dale Forest State Scenic Corridor

The 15-mile stretch of US 395 heading south from Ukiah is an officially designated state scenic highway, and big, old ponderosas are the primary reason why. This picturesque roadway through the Blue Mountains parallels Camas Creek until its confluence with the North Fork of the John Day River. The corridor showcases ponderosas along the creek and on ledges that stairstep their way up the steep canyon walls. The winding highway borders the western edge of the Bridge Creek Wildlife Area, with patches of blackened trees providing evidence of recent wildfire. The area was traditionally used by Native Americans to dig camas roots, hence the creek's name. The scenic corridor lies about 50 miles south of Pendleton via US 395.

Big Pine Campground—Rogue River–Siskiyou National Forest

Oregon not only boasts the largest-diameter ponderosa in the country (Big Red), but also the tallest. Serendipity prevailed when the Big Pine Campground was named before the record-setting tree—Phalanx—was recognized as the tallest pine of any kind in the world.[37] This 268-foot-tall tree and a number of other ponderosas nearly as tall struggle to reach sunlight in a shady old-growth stand dominated by equally large Douglas-fir. Hiking the 0.3-mile Big Pine Loop is the best way to see and appreciate these exceptional trees. Big Pine Campground can be reached by taking the Merlin exit (61) from I-5 about 3 miles north of Grants Pass, driving west for about 12 miles on the Merlin-Galice Road, and then driving southwest on Forest Road 25 (Taylor Creek Road) for another 12 miles to the campground.

Phalanx, the tree in the middle of the photo with a fading crown, is the tallest pine of any kind in the world. It grows west of Grants Pass.

Lost Forest Research Natural Area

Not far from the tiny town of Christmas Valley, a 9,000-acre island of ponderosa pines called the Lost Forest appears out of nowhere in the high desert of south-central Oregon. The Lost Forest features trees up to 600 years old growing among sand dunes in an area that receives much less precipitation than ponderosas normally need to survive. Several factors help explain ponderosa's unexpected appearance in the desert 40 miles from the nearest contiguous forest. An impervious layer of sedimentary rock, known as caliche, lies under the sand and prevents water from draining out of the rooting zone. The coarse sand at the surface allows precipitation to infiltrate quickly and serves as a mulch to reduce evaporation. Finally, Lost Forest ponderosa seeds germinate more quickly than pine seeds elsewhere, a distinct advantage in a droughty environment.[38] It is not clear how the newly germinated seedlings can survive until they are able to tap moisture well below the soil surface, although first-year ponderosa seedlings produce longer taproots than most other trees. Lost Forest can be reached by driving east from Christmas Valley on County 5-14 (Christmas Valley–Wagontire Road), north on County 5-14D (Fossil Lake Road), and then east on County 5-14E (Lost Forest Road).

*Old-growth ponderosa pines growing among
sand dunes in the Lost Forest.*

SOUTH DAKOTA

Peter Norbeck Scenic Byway—Black Hills

The Peter Norbeck Scenic Byway is as much about the highway as it is about the stunning views of ponderosa pines it affords. The byway includes the Wildlife Loop and Needles Highway in Custer State Park, and the Iron Mountain Road (SD 16A), running from the park's east entrance to the town of Keystone. The byway has been listed as one of America's top ten scenic drives by the American Travel Writers. Norbeck, a former US Senator, worked closely with the park superintendent and personally marked the route for the roads. Their ideas, including tunnels to frame scenic vistas and switchback pigtail bridges that allow the road to gain and lose elevation quickly, clashed with all practical principles for road construction. Norbeck also insisted on constructing bridges with massive wood beams, rather than concrete, to complement the signature pine forests. The road turned out to be an engineering masterpiece that engineers said couldn't be built. In Norbeck's words, "This is not meant to be a super highway, to do the scenery justice you should drive no more than 20 miles per hour, and to do it full justice you should simply get out and walk."[39]

Pigtail bridge along Peter Norbeck Scenic Byway amidst ponderosa pines in the Black Hills. —Photo by Brian Koder

Mt. Rushmore National Memorial

A little-known area of old-growth ponderosa pine can be accessed from the concessionaire parking lot at the Mount Rushmore National Memorial. The Blackberry trailhead is located at the edge of the gravel fan just south of the main parking area. It is not well marked, so visitors should inquire at park headquarters for detailed directions. The trail features abundant evidence of past fires in the form of live, fire-scarred old-growth trees and blackened snags and stubs. The trail winds through interesting rock outcrops as it descends into a moist, picturesque canyon less than 1 mile from the memorial. There is also an impressive old-growth stand on the bench just east of the memorial. Bark beetles are active throughout the area, but park managers are attempting to restore resilience to the pines by thinning dense stands and following up with prescribed burning.

Fire-scarred ponderosas along the Blackberry Trail in Mt. Rushmore National Memorial.

TEXAS

Davis Mountains Preserve

Located about 25 miles northwest of Fort Davis via TX 118, this 33,000-acre preserve is a sky island rising above the Chihuahuan Desert of far west Texas. Owned by the Nature Conservancy, the Davis Mountains Preserve features some of the best examples of ponderosa pine in the state, including the largest—a graceful, 43-inch-diameter, 110-foot-tall tree.[40] The Madera Canyon Trail, a 2.5-mile loop, is a good way to see the preserve's ponderosa pine in an authentic west Texas landscape.

Light winter frost on the largest ponderosa pine in Texas, Davis Mountains Preserve. —Photo by Daniel Chamberlin

Guadalupe Mountains National Park

Guadalupe Mountains National Park, a fossil reef jutting high above the surrounding desert, is home to the highest point in Texas—8,749-foot Guadalupe Peak. A good place to see ponderosa pines in the park, which commonly occur intermixed with Douglas-fir, is the bowl area just north of 8,368-foot Hunter Peak.[41] The bowl can be accessed by hiking either the Tejas Trail or the Bear Canyon Trail north from the Pine Springs Visitor Center. The Bowl Trail loop traverses a high plateau through an appealing mosaic of open parks and ponderosas pine–fir forests a half-mile above the desert floor. Guadalupe Mountains National Park is located off US 62/US 180 about 110 miles east of El Paso.

UTAH

Albert Potter's Pines—Somewhere in Utah

In 1902, Albert Potter set out on a daunting, five-month journey to see Utah's forests firsthand. Assigned by Gifford Pinchot, Potter traveled over 3,000 miles by foot, horseback, wagon, and train to evaluate the potential for creating forest reserves in Utah.[42] Potter kept a diary of his travels and mentioned seeing large yellow pines (as ponderosas were referred to) in several locations. Excerpts are included here to entice the sleuth to relocate two of Potter's yellow pine finds.

> August 15—"We struck the old Indian trail and followed it south to Chase Canyon and then on across to Hall's Canyon. . . . The ridge south of Hall's canyon has been . . . cut out by loggers. . . . The next divide is a wide rolling moss-topped one. As we went off of this ridge into the basin of Sixth Water, saw a few yellow pines . . . but only one [big tree] has been spared by the loggers. This tree is 5 ft. D.B.H. [diameter at breast height] and 150 ft. high."

> October 18—"On the South slope of the mountain and on Winslow Creek there is a splendid lot of yellow pines. Many good trees 24 in. D.B.H. and over . . . not averaging over twenty-five or thirty trees per acre. No cutting has been done here. Fires have evidently run through quite frequently. Many of the trees are scar[r]ed."

Wah Wah Mountains

The oldest known living ponderosa pine, a 950-year-old veteran, is found in the Wah Wah Mountains west of Milford in southwestern Utah.[43] The ancient tree recorded only two fires during its nearly 1,000-year lifetime, and both fires occurred during the 1500s. An old ponderosa in an adjacent canyon shows evidence of ten fires. Stan Kitchen, who found the old tree, speculates that Native Americans may have used the other canyon more because of its water source, perhaps explaining the greater fire occurrence there. Getting to the tree—which grows in a BLM Wilderness Study Area—requires a strenuous, three-hour cross-country hike. The ancient tree is located about 100 feet east of the northwest corner of section 32 (T25S, R15W). Interested parties may want to contact the US Forest Service's Shrub Sciences Lab at 735 North 500 East in Provo (801-356-5100) before embarking on a trip to see the Wah Wah pine.

The oldest known living ponderosa pine grows in the Wah Wah Mountains of western Utah. —Photo by Doug Page, Bureau of Land Management

Red Canyon—Dixie National Forest

The drive on UT 12 heading east from US 89 toward Bryce Canyon National Park traverses spectacular Red Canyon at the north end of the Paunsaugunt Plateau. Often referred to as Little Bryce, the Red Canyon corridor features bright green ponderosas set against red-orange to pink sandstone cliffs, spires, and hoodoo formations. Several short trails, including Birdseye, Buckhorn, and Tunnel, as well as a bicycle path paralleling the highway, provide captivating views of pines on the rocks. Red Canyon also boasts a legendary early-day visitor. Butch Cassidy didn't come here for the scenery, but he did come to evade a posse after he thought he had killed a man in a dance-hall fight at nearby Panguitch.[44]

Gnarled, old ponderosa growing near a ridgetop in Red Canyon.

Toms Creek—Deep Creek Mountains

Known as the Deeps, the Deep Creek Mountains are the highest range in western Utah, with two peaks over 12,000 feet. They are also home to what is believed to be the smallest and most isolated population of ponderosa pines in Utah.[45] Numbering only a few hundred, the ponderosas occur as a minor component of a mixed-species stand that also contains Engelmann spruce, subalpine fir, Douglas-fir, limber pine, and aspen. The stand is located in sections 16 and 17, T11S, R18W in the upper Toms Creek drainage of the Deep Creek Mountains. The nearest ponderosas in Utah are about 45 to 50 miles to the southeast. Recent genetics work shows the isolated Toms Creek pines to be genetically similar to those restricted to a small area of southwestern Utah and east-central Nevada. There is no good driving route to the Deeps, so those interested in visiting should contact the Bureau of Land Management's West Desert District office located at 2370 S. Decker Lake Blvd. in Salt Lake City (801-977-4300).

Hells Backbone—Dixie National Forest

In 1933, the Civilian Conservation Corps finished digging and blasting a circuitous route from the eastern edge of the Aquarius Plateau near Boulder across the head of rugged canyons and narrow hogbacks to the town of Escalante, creating a scenic masterpiece called Hells Backbone. Travelers will see ponderosas of myriad shapes and sizes, growing in everything from bare rock to shady, streamside environments. They will also see evidence of thinning, prescribed burning, and wildfire, and lightning-scarred trees. Another highlight is the Hells Backbone Bridge, which spans a 1,500-foot chasm above the confluence of Death Hollow and Sand Creek. Vantage points near the bridge provide spectacular views into the Box–Death Hollow Wilderness. Visitors are encouraged to start the drive from the Boulder end, which is about 1,000 feet higher than Escalante, thereby reducing the number of miles driven uphill on the washboard gravel road. The 35-mile drive leaves UT 12 about 4 miles west of Boulder, first heading northwest, west, and then south to Escalante. Mature ponderosas can be seen growing out of steep, west-facing canyon walls as the road descends the last few miles into Escalante, providing additional evidence of this tree's incredible ability to survive, and even thrive, in extreme environments. The drive south on UT 12 from Torrey to the start of the Backbone drive near Boulder also features photogenic old pines and memorable landscape views to the east.

Hells Backbone Bridge, about midway between Boulder and Escalante.

WASHINGTON

Gifford Pinchot National Forest

The largest ponderosa in Washington grows on the Gifford Pinchot National Forest about 25 miles north of the Columbia River Gorge. The well-formed tree—7 feet in diameter and 202 feet tall—grows among many other large ponderosas rising out of an increasingly dense forest of grand fir and Douglas-fir. It can be reached by driving about 4 miles north from the community of Trout Lake on Forest Road 80, and then turning right (northeast) on Forest Road 8020 for about one-quarter mile to the big tree. As a bonus, this route affords superb views of a giant volcano, Mt. Adams (12,276 feet), about 10 miles to the north.

White Pass Scenic Byway— Mt. Baker–Snoqualmie National Forest

Midway between Yakima and Mt. Rainier, a scenic route over the Cascade crest provides a classic example of a dry-to-moist site vegetation continuum. Known as the White Pass Scenic Byway (WA 12), this route begins about 5 miles west of Naches and heads up the Tieton River Canyon. The treeless, eastern end of the byway starts in sagebrush, but after about 1 mile and a slight increase in elevation, a few scattered ponderosa pines and cottonwoods appear. As the road gradually climbs west into the Cascade Range, larger patches and stringers of trees show up. Soon Douglas-fir and Oregon white oak become part of the mix, and by the 4-mile mark, most of the canyon is covered in forest. Tree density continues to increase for the next few miles, and by about the 8-mile mark forest cover is dense and continuous. Between Rimrock Lake and White Pass, ponderosa pine disappears from the dense forest and is replaced by firs. Ponderosa's occurrence here is similar to other areas on the eastern slope of the Cascades in Washington but different from its typical appearance in the eastern Cascades of Oregon. Ponderosas here typically occur in various mixtures with Douglas-fir, rather than in large, pure stands. Ecologists attribute the greater variation in ponderosa forest composition in Washington to more complex patterns of geology, soils, and topography.

Fort Lewis area near Puget Sound

Ponderosa pine's presence in the moist environment west of the Washington Cascades is surprising enough, but especially in the Fort Lewis vicinity, where it grows close to Puget Sound and just above sea level. Ponderosa established here from 6,000 to 10,000 years ago during a warmer and drier period, when some scientists believe it may have formed a continuous belt from Vancouver,

British Columbia, to southwestern Oregon. Fire suppression since the mid-1800s has allowed the Fort Lewis pines to play the dual ecological role of invaded and invader. Where ponderosa is part of the original forest, lack of periodic fire has allowed Douglas-fir to slowly increase and displace it. Yet this same lack of fire has also allowed ponderosa pines and firs to colonize the formerly treeless prairie. Thinning and prescribed burning treatments are now being used to reduce Douglas-fir in the original forest, and to remove both ponderosa pine and Douglas-fir from the former prairie. Restoration specialist Jeff Foster sees a bright future: "I am excited by the prospect of standing amid a lush cover of native prairie plants, gazing across an all-aged forest of fire-scarred pines to the gleaming dome of Mt. Rainier."[46]

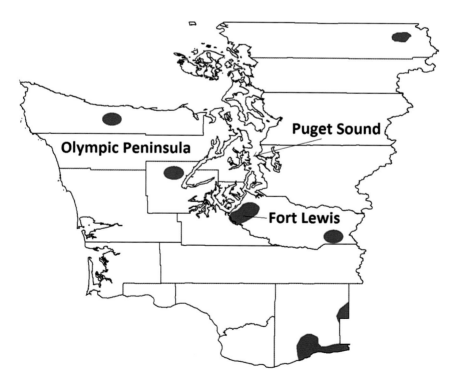

The distribution of ponderosa pine in western Washington. —Map courtesy Kevin Zobrist

Turnbull National Wildlife Refuge

The Turnbull National Wildlife Refuge features a ponderosa pine–bunchgrass forest interspersed with flat rocky areas, rock outcrops, moist meadows, marshes, and potholes. Fire scars on some of the pines provide evidence that surface fires burned through the area prior to active fire suppression in the early 1900s. The refuge's pure ponderosa pine forest is productive habitat for wildlife, with over two hundred species of birds—including twenty-seven species of waterfowl—observed.[47] Thinning and prescribed burning are used to enhance habitat and restore ponderosa forests on parts of the refuge. The Turnbull National Wildlife Refuge is located about 4 miles south of Cheney along the South Cheney-Plaza Road. Refuge headquarters and the start of the 5-mile Pine Creek Auto Tour loop are located about 1 mile east of the same road.

Ponderosa pines growing among potholes in Turnbull National Wildlife Refuge south of Cheney.

Desolation Peak—Ross Lake National Recreation Area

Just south of the Canadian border, a ponderosa pine stand grows in a most unusual place—west of the North Cascades crest. It occurs as an island of dry forest in a sea of moist forest types characteristic of the western Cascades. The ponderosa pine forest is fairly open and features some dry-forest understory species, such as balsamroot.[48] It is located near Ross Lake on the lower slopes of Desolation Peak at an elevation of about 2,000 feet. The stand can be accessed using the Ross Lake boat taxi service that provides trips to the base of Desolation Peak. Noted author Jack Kerouac spent the summer of 1956 at the Desolation Peak fire lookout and later wrote about it in the novel *Desolation Angels*.[49]

Leavenworth Loop—Wenatchee National Forest

The Leavenworth Loop Drive travels north from Leavenworth to Lake Wenatchee on the Chumstick Highway and returns via US 2. Ponderosa pine may seem out of place in an area that promotes its Bavarian look, but the Chumstick Highway features more relatively pure pine forests than can be found in most of Washington's eastern Cascades. The curvy, northern half of this highway has both even- and uneven-aged stands, with some striking, older ponderosas growing on dry rock outcrops. The return part of the loop via US 2 supports fewer pine, typically intermixed with Douglas-fir. One sidelight of this scenic return route along the Wenatchee River is the Tumwater Botanical Area, which boasts the only place in the world where a rare wildflower, showy stickseed, can be found. May or June is the best time to drive this route to catch other showy wildflowers, such as Tweedy's lewisia, yellow-white larkspur, and Thompson's clover. Note: The Tumwater Botanical Area is no longer a formally recognized botanical area, but its location is shown on older Wenatchee National Forest maps.

WYOMING

Blacks Fork River—Mountain View

Even a lifelong ponderosa pine aficionado would not expect to find pines growing in conditions that resemble interior Alaska. However, ponderosas can be found in just such conditions along the Blacks Fork River southwest of Mountain View. Here, scattered ponderosas grow with stunted, and occasionally larger, Engelmann spruce in floodplain areas that resemble the black spruce and muskeg of the Far North.[50] This isolated population of pine—growing at nearly 7,500 feet elevation—was not included in Little's (1971) authoritative map of ponderosa pine distribution. At a distance, the Blacks Fork spruce stands form the distinctive spiked skyline associated with spruce communities of the Far North; up close, one sees they are sprinkled with juniper and an occasional ponderosa pine. Visitors should take exit 39 from I-80 (about 35 miles east of Evanston), proceed south on WY 414 to Mountain View, and then follow WY 410 south and west past Robertson to the Blacks Fork River bridge and about 1 mile beyond.

The forested floodplain along the Blacks Fork River resembles the spiked skyline formed by spruce forests of the Far North. Upon approach, an occasional ponderosa pine can be observed among the Engelmann spruce.

Vedauwoo—Medicine Bow National Forest

Vedauwoo (pronounced *VEE da voo*) is a popular rock climbing, hiking, and camping destination located just 10 miles southeast of Laramie. The Arapaho Indians historically used Vedauwoo and attributed the whimsical shapes and arrangement of the area's granite hoodoos and outcrops to playful spirits. Although most ponderosas growing at Vedauwoo exhibit two or three needles per fascicle (cluster), typical of the ponderosa pine variety *scopulorum*, an occasional four-needle fascicle is also found. Vedauwoo was the only location out of 104 intensively sampled sites in a range-wide study of ponderosa pines where four-needle fascicles were found.[51] Author Carl Fiedler sampled ten fascicles per tree on one hundred pines scattered across Vedauwoo and found only one four-needle fascicle. However, BLM forester Bob Means reports a higher proportion of four-needle fascicles at the eastern edge of Vedauwoo, near the national forest boundary.[52] The collective experience suggests that looking for four-needle fascicles is somewhat akin to searching for a four-leaf clover, only with somewhat better odds. Another

This 2-foot tall ponderosa pine growing on a large rock at Vedauwoo is bearing about a half-dozen cones.

anomaly found here was a 2-foot tall pine with a developing cone crop, the shortest ponderosa pine known to bear cones. To access the Vedauwoo area, visitors can take exit 323 from I-80 and then drive southeast on WY 210 (Happy Jack Road).

Mallo Camp—Black Hills

The Mallo Camp area is located on BLM lands about 20 miles north of Newcastle, adjacent to the South Dakota border. The Mallo Camp drainage is only accessible by trail and contains some of the last untouched old-growth ponderosa pine in Wyoming's Black Hills. The oldest tree in the stand dates back to 1641, with most trees originating around 1735.[53] The BLM has conducted aggressive selection cutting treatments around the stand to reduce tree density and increase resistance to bark beetle attack. The opposite side of the Mallo Camp drainage supports white spruce, which is near the southern edge of this species' range in the western United States. Directions to the Mallo Camp area can be obtained by contacting the BLM Field Office in Newcastle located at 1101 Washington Boulevard (307-746-6600).

Pine Bluffs Rest Area

Located along the old Texas Cattle Trail, Pine Bluffs can be explored from the Pine Bluffs Rest Area by taking exit 401 from I-80, which is the last exit from the interstate before entering Nebraska. This forested bluff in the middle of the prairie was heavily used by Native Americans, and the University of Wyoming maintains an archaeological site here that is staffed by students during the summer months. The Pine Bluffs Rest Area also features an interpretive center and a network of short trails through ponderosa pine, limber pine, and Rocky Mountain juniper stands. The ponderosas here are many-aged, including some centuries old, platy-barked veterans clinging to the edge of the bluffs. The limber pines in the Pine Bluffs area represent the easternmost occurrence of this species in the prairie, as occurrences in North Dakota and South Dakota are in woodland or forested country.[54]

*A weather-beaten pine living on the edge
in the Pine Bluffs area east of Cheyenne.*

BRITISH COLUMBIA

Hanging Tree—Lillooet

Justice was harshly served during the rough-and-tumble era of the Fraser River gold rush, often by hanging the accused from the end of a rope. During the late 1850s, twelve men were purportedly hung from a gnarled old ponderosa pine known as the Lillooet Hanging Tree. Entries in the presiding judge's diaries show that he sentenced several men accused of stealing cattle to hang in order to keep the peace, while the other victims were likely common thieves hung by frontier justice.[55] As one old-timer put it, "If you were gonna hold a hangin', that's for sure a tree you'd think about usin'—and a damn pretty view for the one bein' hanged to see."[56] The Hanging Tree died years ago, but its skeleton still stands in Cayoosh Park overlooking the town of Lillooet and the Fraser River, about 70 miles west of Kamloops.

Anderson Lake area—Squamish Forest District

Ponderosa pine appears in a most unlikely part of British Columbia—the Anderson Lake area north of Vancouver. Here, ponderosas up to 3 feet in diameter and 400 years old grow among such moist-site species as coastal Douglas-fir, western redcedar, western hemlock, and paper birch. Ponderosa's ability to maintain a foothold on these moist sites can be attributed to forest burning by First Nations (Salish people), who have lived in the area for over 8,000 years.[57] Dating back to the 1500s, the historical average fire interval in the area was six years. Indians used fire to rejuvenate berry patches, maintain meadows for production of edible roots, top-kill hazelnut to increase yields, and stimulate big-game forage plants. Ethnobotanists estimate that the Salish used about 35 to 40 gallons of berries and over 200 pounds of roots per family per year. Given that the Interior Salish population in 1835 was estimated at 13,500, maintaining food production areas in these moist forests required continual burning, and ponderosa pine was an indirect beneficiary of the historical burning regime. Burning stopped about a century ago, and there has been no pine regeneration in the area for decades. Restoration treatments starting in 1998 were aimed at regenerating ponderosa pine and reinvigorating the old pines.[58] However, the

Relict old ponderosas remaining after restoration cutting and burning in a moist forest north of Vancouver. The white seedling protectors safeguard the Douglas-fir seedlings that were planted after the management objective for the area changed.
—Photo by Robert Gray

effort was terminated in 2006 when the government refocused on timber production. Except for one hillside along lower Haylmore Creek, all treated areas have been planted with Douglas-fir seedlings. Directions to the restoration area can be obtained by contacting the BC Ministry of Forests District Office at 42000 Loggers Lane in Squamish (604-898-2100).

Rocky Mountain Trench Ecosystem Restoration Program

The most ambitious open forest and grassland restoration effort in British Columbia extends from Roosville at the US border near Eureka, Montana, north for 150 miles to Radium Hot Springs. Partners in the Rocky Mountain Trench restoration effort have restored open forest and livestock-wildlife range on about 125,000 acres of private land, Crown (public) land, First Nations reserves, and provincial parks since 1998.[59] Grasslands and open forests constitute less than 10 percent of the Rocky Mountain Trench landscape, yet nearly 70 percent of the area's rare and endangered species depend on these communities during all or part of their life cycles. Animals important to people, including elk, bighorn sheep, and livestock, are also primary beneficiaries of restored open forest and range habitats. Recent restoration treatments can be seen along the east side of BC 93 just south of Radium Hot Springs. Visitors will notice that treated areas here, which are near the northern range limits of ponderosa pine, are more open than restored ponderosa forests in the United States. Of interest to birders, the 11,000-acre Dutch-Findley restoration project between Canal Flats and Fairmont Hot Springs was designed to create preferred habitat for the rare Lewis's woodpecker, including open forest for foraging, abundant snags for nesting, and shrubby undergrowth for berry production. This is the only woodpecker that catches insects on the fly rather than plucking them from trees, and it also feeds on berries from native shrubs. More information about Rocky Mountain Trench restoration projects and plans can be found at www.trench-er.com.

Ponderosa pines grow among hybrid spruce in the floodplain at Canal Flats, British Columbia.

Scientific Names of Plants and Animals Mentioned in Text

COMMON NAME	SCIENTIFIC NAME

Trees

Birch	*Betula* spp.
Black cottonwood	*Populus trichocarpa*
Black spruce	*Picea mariana*
Cottonwoods	*Populus* spp.
Douglas-fir (coastal)	*Pseudotsuga menziesii* var. *menziesii*
Douglas-fir (inland)	*Pseudotsuga menziesii* var. *glauca*
Engelmann spruce	*Picea engelmannii*
Gambel oak	*Quercus gambelii*
Giant sequoia	*Sequoiadendron gigantea*
Grand fir	*Abies grandis*
Junipers	*Juniperus* spp.
Mountain mahogany	*Cercocarpus* spp.
Oregon white oak	*Quercus garryana*
Paper birch	*Betula papyrifera*
Pines	
Apache pine	*Pinus engelmannii*
Arizona pine	*Pinus arizonica*
Border pinyon	*Pinus discolor*
Bristlecone pine	*Pinus longaeva*
Chihuahuan pine	*Pinus leiophylla*
Jeffrey pine	*Pinus jeffreyi*
Limber pine	*Pinus flexilis*
Lodgepole pine	*Pinus contorta*
Longleaf pine	*Pinus palustris*
Pinyon pine	*Pinus edulis*
Ponderosa pine (yellow pine)	*Pinus ponderosa*
Red pine	*Pinus resinosa*
Singleleaf pinyon	*Pinus monophylla*
Sugar pine	*Pinus lambertiana*

219

Washoe pine	*Pinus ponderosa* var. *washoensis*
Western white pine	*Pinus monticola*
Whitebark pine	*Pinus albicaulis*
Quaking aspen	*Populus tremuloides*
Redwood	*Sequoia sempervirens*
Redwood (prehistoric)	*Sequoia affinis*
Rocky Mountain juniper	*Juniperus scopulorum*
Subalpine fir	*Abies lasiocarpa*
Western hemlock	*Tsuga heterophylla*
Western larch	*Larix occidentalis*
Western redcedar	*Thuja plicata*
White fir	*Abies concolor*
White spruce	*Picea glauca*

OTHER PLANTS

Balsamroot	*Balsamorhiza* spp.
Bunchgrass	several species of different genera, including *Festuca* and *Agropyron*
Camas	*Camassia* spp.
Hazelnut	*Corylus cornuta*
Dwarf mistletoe	*Arceuthobium* spp.
Kinnikinnick	*Arctostaphylos uva-ursi*
Palmetto	*Serenoa repens*
Pinegrass	*Calamagrostis rubescens*
Showy stickseed	*Hackelia venusta*
Thompson's clover	*Trifolium thompsonii*
Tweedy's lewisia	*Lewisia tweedyi*
Willows	*Salix* spp.
Yellow-white larkspur	*Delphinium xantholeucum*

BIRDS

Blue grouse	*Dendragapus obscurus*
Bobwhite quail	*Colinus virginianus*
Clark's nutcracker	*Nucifraga columbiana*
Flammulated owl	*Otus flammeolus*
Golden eagle	*Aquila chrysaetos*
Goshawk	*Accipiter gentilis*
Great gray owl	*Strix nebulosa*
Great horned owl	*Bubo virginianus*
Lewis's woodpecker	*Melanerpes lewis*
Mexican spotted owl	*Strix occidentalis lucida*
Mountain bluebird	*Sialia currucoides*
Northern flicker	*Colaptes auratus*
Northern spotted owl	*Strix occidentalis caurina*
Pileated woodpecker	*Dryocopus pileatus*
Red crossbill	*Loxia curvirostra*
Wild turkey (Merriam's turkey)	*Meleagris gallopavo*
Woodpeckers	*Picoides* spp.

MAMMALS

Abert's squirrel (tassel-eared squirrel)	*Sciurus aberti*
Antelope (pronghorn)	*Antilocapra americana*
Bighorn sheep	*Ovis canadensis*
Black bear	*Ursus americanus*
Chipmunk	*Tamias* spp.
Coyote	*Canis latrans*
Deer	*Odocoileus* spp.
Deer mouse	*Peromyscus maniculatus*
Elk	*Cervus canadensis*
Flying squirrel	*Glaucomys sabrinus*
Fox	*Vulpes* spp.
Mule deer	*Odocoileus hemionus*
Raccoon	*Procyon lotor*
Tree squirrel	*Sciurus* spp.
Wild horse	*Equus ferus*

INSECTS AND FISH

Engraver beetle	*Ips* spp.
Mountain pine beetle	*Dendroctonus ponderosae*
Roundheaded pine beetle	*Dendroctonus adjunctus*
Western pine beetle	*Dendroctonus brevicomis*
Apache trout	*Oncorhynchus gilae apache*
Gila trout	*Oncorhynchus gilae*

NOTES

Preface

1. Phelps, Autobiography with Letters, p. 965.
2. Powell, *The Exploration of the Colorado River and Its Canyons*, Preface.
3. Worster and others, *"The Legacy of Conquest*: A Panel of Appraisal," p. 303.
4. Maclean, *A River Runs Through It and Other Stories*, p. 101–2.

Part I

Chapter 1. A Tale of Two Forests

1. Evans, *Powerful Rockey: The Blue Mountains and the Oregon Trail*, p. 238.
2. Cooper, "Changes in vegetation, structure, and growth of southwestern pine forests since white settlement."
3. Hammer, Stewart, and Radeloff, "Demographic trends, the wildland-urban interface, and wildfire management."
4. Arno and Allison-Bunnell, *Flames in Our Forest: Disaster or Renewal?*, p. 124.

Chapter 2: Indians in the Pines

1. Urbaniak, *Anasazi of Southwest Utah: The Dance of Shadow and Light*, p. 76.
2. D. Ford, Chaco Culture National Historic Park, personal communication, 2013.
3. English and others, "Strontium isotopes reveal distant sources of architectural timbers in Chaco Canyon, New Mexico."
4. Reynolds and others, "$^{87}Sr/^{86}Sr$ sourcing of ponderosa pine used in Anasazi great house construction at Chaco Canyon, New Mexico."
5. Murphy, *Graced by Pines: The Ponderosa Pine in the American West*, p. 39–40.
6. Simpson, *Navaho Expedition: Journal of a Military Reconnaissance*.
7. Lister and Lister, *Chaco Canyon: Archaeology and Archaeologists*.
8. Nash, *Time, Trees, and Prehistory: Tree-Ring Dating and the Development of North American Archaeology 1914–1950*.
9. Speer, *Fundamentals of Tree Ring Research*, p. 153–54.
10. Douglass, "The secret of the Southwest solved by talkative tree rings."
11. D. Ford, Chaco Culture National Historic Park. Personal communication.
12. Ibid.
13. Hualapai Tribe, http://hualapai-nsn.gov/.

14. Loosle, "Ponderosa bark used for food, glue, and healing."
15. Standley, "Some useful native plants of New Mexico."
16. White, *Scarred Trees in Western Montana*.
17. Swetnam, "Peeled ponderosa pine trees: A record of inner bark utilization by Native Americans."
18. Kingsbury and Dixon, "American Indian culturally modified ponderosa pine trees on the Payette National Forest."
19. Willey, *Nez Perce News*, June 9, 1881.
20. Moulton, *The Journals of the Lewis and Clark Expedition*.
21. Ibid.
22. Great Sand Dunes National Park, http://www.nps.gov/grsa/historyculture/index.htm.
23. Oregon Travel Experience, Indian Village Grove, http://ortravelexperience.com/oregon-heritage-trees/indianvillage-grove/.
24. Moulton, *The Journals of the Lewis and Clark Expedition*.
25. Anderton, McAvoy, and Kuhns, *Native American Uses of Utah Forest Trees*; Vestal, "The ethnobotany of the Ramah Navaho."
26. Palmer, "Shuswap Indian ethnobotany"; Anderton, McAvoy, and Kuhns, *Native American Uses of Utah Forest Trees*.
27. Hart, *Montana Native Plants and Early Peoples*, p. 51.
28. Palmer, "Shuswap Indian ethnobotany."
29. Hart, "The ethnobotany of the Northern Cheyenne Indians of Montana"; Anderton, McAvoy, and Kuhns, *Native American Uses of Utah Forest Trees*.
30. Mahar, *Ethnobotany of the Oregon Paiutes of the Warm Springs Indian Reservation*; Anderton, McAvoy, and Kuhns, *Native American Uses of Utah Forest Trees*.
31. Hart, "The ethnobotany of the Northern Cheyenne Indians of Montana"; Anderton, McAvoy, and Kuhns, *Native American Uses of Utah Forest Trees*.
32. Hart, *Montana Native Plants and Early Peoples*, p. 51; Mahar, *Ethnobotany of the Oregon Paiutes of the Warm Springs Indian Reservation*.
33. Moulton, *The Journals of the Lewis and Clark Expedition*.
34. Matthews, *Montana Main Streets: Vol. 6. A Guide to Historic Missoula*.
35. Winkler, typed transcript of oral history interview, on file at Idaho State Historical Society.

Chapter 3. Pioneers in the Pines

1. Covey, *Cabeza de Vaca's Adventures in the Unknown Interior of America*.
2. Winship, "The Coronado Expedition, 1540–1542," p. 495, 511; A. Clark-Sanchez, National Park Service, personal communication.
3. Abarr, "Inscription Rock."
4. Museum of New Mexico Foundation, New Mexico History Museum, http://www.museumfoundation.org/museums/new-mexico-history-museum-palace-of-the-governors.
5. Santa Fe Unlimited, The Historic Churches of Santa Fe, http://churches-sf2.nm-unlimited.net/.

6. Hunt, "The beginning of the Lewis and Clark Expedition."

7. Moulton, *The Journals of the Lewis and Clark Expedition*.

8. Ibid.

9. Ibid.

10. Geyer, "Notes on the vegetation and general character of the Missouri and Oregon Territories," p. 204.

11. Moulton, *The Journals of the Lewis and Clark Expedition*.

12. Ibid.

13. Callaham, Pinus ponderosa: *A Taxonomic Review with Five Subspecies in the United States*.

14. Palladinos, *Anthony Ravalli, S.J.: Forty Years a Missionary in the Rocky Mountains*.

15. Fanselow, *Traveling the Oregon Trail*.

16. Crawford, *Journal of Medorem Crawford*, p 18.

17. Evans, *Powerful Rockey: The Blue Mountains and the Oregon Trail*.

18. Frémont, *Memoirs of My Life*.

19. Shuford, "Logging in Shasta County."

20. Cunningham, "Famous old sawmills of the Shingletown Ridge"; J. M. Tuggle and J. Montgomery. Shasta Historical Society, personal communication, 2014.

21. Hoover and others, *Historic Spots in California*.

22. California Pioneer Heritage Foundation, http://californiapioneer.org/destinations /emigrant-trail.

23. California Department of Finance, The Gold Rush Plants the Seed, http://www.dof .ca.gov/html/fs_data/historycaeconomy/gold_rush.htm.

24. Galloway, *The First Transcontinental Railroad: Central Pacific, Union Pacific*.

25. Central Pacific Railroad Photographic History Museum, History, http://cprr.org /Museum/Newspapers/.

26. Smith, "The historic Blue Ridge Flume of Shasta and Tehama Counties, California."

27. Hutchinson, *California Heritage: A History of Northern California Lumbering*; Smith, "The historic Blue Ridge Flume."

28. Hutchinson, *California Heritage*.

29. *Weekly Sentinel* (Red Bluff), August 1, 1874; Smith, "The historic Blue Ridge Flume."

30. Lewis, *History of Victorian Tehama County, California*; Smith, "The historic Blue Ridge Flume."

31. Hutchinson, *California Heritage*.

32. Ibid.

33. Johnson, *Chips and Sawdust*.

34. Olberding, *"It Was a Young Man's Life": G. A. Pearson*.

35. Woolsey, *Western Yellow Pine in Arizona and New Mexico*.

36. Wallace, *The Great Reconnaissance: Soldiers, Artists, and Scientists on the Frontier 1848–1861*.

37. Reed, *The Puebloan Society of Chaco Canyon*.

38. Simpson, *Navaho Expedition: Journal of a Military Reconnaissance*, p. 73.

39. Whipple, *Reports of Explorations and Surveys*, Part I, p. 79, 82.

40. Ibid.

41. Gray, Farquhar, and Lewis, "Camels in western America."

42. Powell, *The Exploration of the Colorado River and Its Canyons*, p. 259; Leavengood, "History of Phantom Ranch."

43. *St. Paul Pioneer*. July 3, 1874; Krause and Olson, *Prelude to Glory*.

44. *New York World*. August 16, 1874; Krause and Olson, *Prelude to Glory*.

45. *Chicago Inter-Ocean, New York World*, and *St. Paul Pioneer*, multiple reports between June 24 and September 9, 1874; Krause and Olson, *Prelude to Glory*.

46. Horsted, *The Black Hills Yesterday and Today*.

47. *Yankton Press*, March 13, 1872; Jackson, *Custer's Gold*.

48. *Yankton Press*, March 20, 1872; Jackson, *Custer's Gold*.

49. South Dakota State Historical Society and South Dakota Department of History, *South Dakota Historical Collections*, vol. 7, p. 590.

50. DeQuille, *History of the Big Bonanza*, p. 238, 240.

51. Ibid.

52. Kahn, "Flume with a view: Saving an American engineering marvel."

53. Ibid.

54. Anaconda Forest Products Company Records, Maureen and Mike Mansfield Library.

55. National Park Service, Discover History, Cable Mountain Draw Works, http://www.nps.gov/hfc/pdf/sitebulletin/cable_mtn_bl.pdf.

56. Ibid.

57. Skagit River Journal, online at http://skagitriverjournal.com/Upriver/Uto-Conc/Lyman/Klement/Klement09-CascadePass.html.

58. Halliday, "The forgotten father of North Cascades National Park."

Chapter 4. A Special Tree

1. Haller and Vivrette, "Ponderosa pine revisited."

2. Burns and Honkala, *Silvics of North America*, vol. 1.

3. Callaham, Pinus ponderosa: *A Taxonomic Review with Five Subspecies in the United States*.

4. Wellner, *Frontiers of Forestry Research—Priest River Experimental Forest, 1911–1976*; DeWald and Mahalovich, "Historical and contemporary lessons from ponderosa pine genetic studies at the Fort Valley Experimental Forest, Arizona."

5. Vander Wall and Balda, "Coadaptations of the Clark's nutcracker and the piñon pine for efficient seed harvest and dispersal."

6. Arno, *Timberline: Arctic and Alpine Forest Frontiers*.

7. Bates, *Forest Types of the Central Rocky Mountains as Affected by Climate and Soil*.

8. Burns and Honkala, *Silvics of North America*.

9. Graham and Jain, *Proceedings of the Symposium on Ponderosa Pine*.

10. Mills, *The Story of a Thousand-Year Pine*.

11. *Montana Record-Herald*, "Giant yellow pine 1,100 years old is felled near Evaro," January 19, 1926.

12. Van Pelt, *Forest Giants of the Pacific Coast*.

Chapter 5. Forests Born of Fire

1. Swetnam, "Fire history and climate change in giant sequoia groves."
2. Cooper, "Report on the botany of the route," p. 10.
3. Arno and Allison-Bunnell, *Flames in Our Forest: Disaster or Renewal?*
4. Powell, "Testimony to Congress," p. 207–8.
5. Boyd, *Indians, Fire, and the Land in the Pacific Northwest*; Stewart, *Forgotten Fires: Native Americans and the Transient Wilderness*; Anderson, *Tending the Wild: Native American Knowledge and the Management of California's Natural Resources*; Salish–Pend d'Oreille Cultural Committee, *The Salish People and the Lewis and Clark Expedition*.
6. Moulton, *The Journals of the Lewis and Clark Expedition*, p. 179.
7. Biondi, "Decadal-scale dynamics at the Gus Pearson Natural Area: Evidence for inverse (a)symmetric competition?"

Chapter 6. Crusaders against Fire

1. Arno and Allison-Bunnell, *Flames in Our Forest: Disaster or Renewal?*
2. Pyne, *Fire in America: A Cultural History of Wildland and Rural Fire.*
3. Coe, *Frontier Doctor*, p. 237–38.
4. McKay, *Trails of the Past: Historical Overview of the Flathead National Forest, 1800–1960.*
5. Pyne, *Fire in America.*
6. Pyne, *Year of the Fires: The Story of the Great Fires of 1910*; Egan, *The Big Burn.*
7. Ibid.
8. Spencer, *The Big Blow-Up*, p. 80.
9. Pyne, *Year of the Fires.*
10. Pyne, *Fire in America.*
11. Pinchot, "The relation of forests and forest fires."
12. Holtmeier, "Geoecological aspects of timberlines in northern and central Europe."
13. Leiberg, "The Bitterroot Forest Reserve," p. 391–92.
14. Pyne, *Fire in America*; Lutts, "The trouble with Bambi: Walt Disney's Bambi and the American vision of nature."
15. Gruell, *Fire and Vegetative Trends in the Northern Rockies: Interpretations from 1871–1982 Photographs*; Smith and Arno, *Eighty-Eight Years of Change in a Managed Ponderosa Pine Forest*; Gruell, *Fire in Sierra Nevada Forests: A Photographic Interpretation of Ecological Change since 1849.*
16. Swetnam and Baisan, "Historical fire regime patterns in the southwestern United States since 1700."

Chapter 7. Advocates for Burning

1. Pinchot, *Breaking New Ground*; Pyne, *Fire in America*, p. 102.
2. Barrett, *A Record of Forest and Field Fires in California*, cited in Pyne, *Fire in America,* p. 102.
3. Quoted by Biswell, "Effects of fire on chaparral," p. 330.

4. Pyne, *Fire in America*; Arno and Allison-Bunnell, *Flames in Our Forest*.

5. Ibid.; Robertson, *Ecosystem Management of the National Forests and Grasslands*.

6. Pyne, *Fire in America*; Holtmeier, "Geoecological aspects of timberlines in northern and central Europe."

7. Earley, *Looking for Longleaf: The Fall and Rise of an American Forest*, p. 13.

8. Chapman, *Factors Determining Natural Reproduction of Longleaf Pine*; Schiff, *Fire and Water*; Pyne, *Fire in America*.

9. *New York Times*, August 26, 1910, p. 4; Pyne, *Fire in America*.

10. Hoxie, "How fire helps forestry."

11. Pyne, *Fire in America*; Pyne, *Year of the Fires*.

12. Kitts, "Preventing forest fires by burning litter."

13. White, "Woodsmen, spare those trees!"; Graves, "The torch in the timber."

14. Pyne, *Fire in America*.

15. Ibid.; Schiff, *Fire and Water*.

16. Greene, "The forest that fire made."

17. Stoddard, *The Bobwhite Quail: Its Habits, Preservation, and Increase*.

18. Pyne, *Fire in America*, p. 115.

19. Ibid., p.113.

20. Ibid., p. 116; Schiff, *Fire and Water*, p. 98–100.

21. Weaver, "Fire as an ecological and silvicultural factor in the ponderosa pine region of the Pacific Slope"; Carle, *Burning Questions: America's Fight with Nature's Fire*.

22. Weaver, "Fire and its relationship to ponderosa pine," p.128.

23. Ibid.

24. Weaver, "Fire as an ecological and silvicultural factor," p. 7.

25. Carle, *Burning Questions*, p.62.

26. Ibid., p. 62–63.

27. Ibid., p. 57.

28. Ibid., p. 67–69.

29. Nelson, "Fire management policy in the national forests: A new era."

30. Koch, "The passing of the Lolo Trail."

31. Carson, *Silent Spring*; Leopold and others, "Wildlife management in the national parks."

32. Pyne, *Fire in America*; Odum, *Ecology and Our Endangered Life-Support Systems*.

Chapter 8. Logging Legacy— From Clearfelling to Clearcutting

1. Schubert, *Silviculture of Southwestern Ponderosa Pine*.

2. Graves, "The Black Hills Forest Reserve"; Symstad and Bynum, *The Extent and Significance of Old-Growth Ponderosa Pine Forest at Mount Rushmore National Memorial*.

3. Hoffman and Krueger, "Forestry in the Black Hills."

4. Koch, *Forty Years a Forester, 1903–1943*, p. 71–73.

5. Ibid.

6. Howe, "Genetic effects of uneven-aged management."

7. Baker and others, *Timeless Heritage: A History of the Forest Service in the Southwest.*

8. Pearson, *Management of Ponderosa Pine in the Southwest.*

9. Pearson, *Natural Reproduction of Western Yellow Pine in the Southwest.*

10. Olberding, *"It Was a Young Man's Life": G. A. Pearson.*

11. Lang and Stewart, *Reconnaissance of the Kaibab National Forest.*

12. Garrett, Soulen, and Ellenwood, "After 100 years of forest management: The North Kaibab."

13. Joslin, *Ponderosa Promise: A History of US Forest Service Research in Central Oregon.*

14. Ibid.

15. Ibid.

16. Oregon Historical Society, Oregon History Project, narratives, http://www.ohs.org /education/oregonhistory/narratives/subtopic.cfm?subtopic_ID=401.

17. Ibid.

18. Joslin, *Ponderosa Promise.*

19. Munger, *Western Yellow Pine in Oregon.*

20. Joslin, *Ponderosa Promise.*

21. Show and Kotok, *The Role of Fire in the California Pine Forests.*

22. Anderson, *Application of Selective Logging to a Ponderosa Pine Operation in Western Montana.*

23. Meyer, *Growth in Selectively Cut Ponderosa Pine Forests of the Pacific Northwest.*

24. Pearson, *Natural Reproduction of Western Yellow Pine in the Southwest.*

25. Swanson, *The Bitterroot and Mr. Brandborg.*

26. Steen, *The US Forest Service: A History.*

27. Backus, "Bitterroot National Forest: Officials seek options for thinning." *Missoulian* July 8, 2009.

28. Steen, *The US Forest Service: A History.*

29. Ibid.

30. Backus, "Bitterroot National Forest: Officials seek options for thinning." *Missoulian* July 8, 2009.

31. Ibid.

32. Swanson, *The Bitterroot and Mr. Brandborg.*

33. Burk, *Clearcut Crisis: Controversy in the Bitterroot.*

34. Swanson, *The Bitterroot and Mr. Brandborg.*

35. Ibid.

36. US Congress, *Multiple-Use Sustained Yield Act of 1960*; Burke, *Clearcut Crisis*, p. 63.

37. W. O. Potter, personal communication, 2003.

38. Bolle and others, *A University View of the Forest Service.*

39. Ibid.

40. Ibid.

Chapter 9. Loving the Forests to Death

1. Thomas, "Dynamic vs. static management in a fire-influenced landscape—The Northwest Forest Plan."

2. Worster, *Nature's Economy: A History of Ecological Ideas.*

3. Rogers, *Disturbance Ecology and Forest Management.*

4. Agee, "The fallacy of passive management of western forest reserves," p. 19.

5. Arno and Fiedler, *Mimicking Nature's Fire: Restoring Fire-Prone Forests in the West.*

6. US Fish and Wildlife Service, *Revised Recovery Plan for the Northern Spotted Owl.*

7. Franklin and others, "Old-growth conifer forests."

8. Swanson, *The Bitterroot and Mr. Brandborg,* chapter 8.

9. Clary, *Timber and the Forest Service;* Hirt, *A Conspiracy of Optimism: Management of the National Forests since World War Two.*

10. Biswell and others, *Ponderosa Fire Management;* Kilgore and Curtis, *Guide to Understory Burning in Ponderosa Pine–Larch–Fir Forests in the Intermountain West.*

11. Ibid.; Clary, *Timber and the Forest Service;* Hirt, *A Conspiracy of Optimism.*

12. DeBonis, "20 years of activism."

13. Mortimer and Malmsheimer, "The Equal Access to Justice Act and US Forest Service land management: Incentives to litigate?"; Baier, "Reforming the Equal Access to Justice Act."

14. *Missoulian,* Op-ed, March 19, 2007; see also Hillis, "Tweak laws to keep fringe from environmental lawsuit frenzy," *Missoulian,* April 8, 2007.

15. Headwaters Economics, "Solutions to the rising costs of fighting fires in the wildland-urban interface."

16. Radeloff and others, "Housing growth in and near United States protected areas limits their conservation value."

Chapter 10. Forests under Siege— From Megafires to Bark Beetles

1. B. Armstrong, Santa Fe National Forest, personal communication, 2014.

2. Leopold, *A Sand County Almanac and Sketches Here and There,* p. 133.

3. Fulé and others, "Mixed-severity fire regime in a high-elevation forest of Grand Canyon, Arizona, USA."

4. Brown, Wienk, and Symstead, "Fire and forest history at Mount Rushmore."

5. Pierce, Meyer, and Jull, "Fire-induced erosion and millennial-scale climate change in northern ponderosa pine forests."

6. Brown, Kaufman, and Shepperd, "Long-term, landscape patterns of past fire events in a montane ponderosa pine forest of central Colorado"; Sherriff and Veblen, "A spatially-explicit reconstruction of historical fire occurrence in the ponderosa pine zone of the Colorado Front Range."

7. Iniguez, Swetnam, and Baisan, "Spatially and temporally variable fire regime on Rincon Peak, Arizona."

8. Cooper, "Changes in vegetation, structure, and growth of southwestern pine forests since white settlement."

9. Ibid.

10. Swetnam and Baisan, "Historical fire regime patterns in the southwestern United States since 1700."

11. Westerling and others, "Warming and earlier spring increase western US forest wildfire activity."

12. Brown, Hall, and Westerling, "The impact of twenty-first-century climate change on wildland fire danger in the western United States."

13. Peterson, "Good policy and good intentions won't stop big wildfires."

14. Trachtman, "Fire fight."

15. Vosick, *An Era of Megafires.*

16. J. A. Youtz, US Forest Service, Southwestern Region, personal communication, 2013.

17. Vosick, *An Era of Megafires.*

18. Ibid.

19. Ibid.

20. J. A. Youtz, US Forest Service, Southwestern Region, personal communication, 2013.

21. Ibid.

22. Tidwell, Statement before the Committee on Energy and Natural Resources.

23. Vosick, *An Era of Megafires.*

24. P. Brown, Rocky Mountain Tree-ring Research, personal communication, 2015.

25. Vosick, *An Era of Megafires.*

26. J. A. Youtz, US Forest Service, Southwestern Region, personal communication, 2014.

27. S. Coleman, Apache-Sitgreaves National Forests, Alpine Ranger District, personal communication, 2013.

28. Ibid.

29. Peterson, "Good policy and good intentions won't stop big wildfires."

30. Western Forestry Leadership Coalition, *The True Cost of Wildfire in the Western US.*

31. Carlson, Dobrowski, and Safford, *Angora Fire Vegetation Monitoring Annual Progress Report.*

32. Marshall, "The high cost of fire."

33. Ibid.

34. Fiedler, Dodson, and Metlen, "Exotic plant response to forest disturbance in the western United States."

35. R. Steinke, Coconino National Forest, personal communication, 2013.

36. Watershed Research and Education Program. http://nau.edu/Watershed-Research -Education/.

37. Fettig and others, "The effectiveness of vegetation management practices for prevention and control of bark beetle infestations in coniferous forests of the western and southern United States."

38. Ball, "Forest health conditions and forest management practices on the Black Hills National Forest."

39. Ibid.

40. Weaver, "Fire as an ecological and silvicultural factor in the ponderosa pine region of the Pacific Slope."

41. C. J. Fettig, Pacific Southwest Research Station, personal communication, 2013.

42. Ibid.

43. Ibid.

44. Ibid.

45. Ibid.

Chapter 11. Restoration—Is It Too Late?

1. C. Pileski, Eastern Montana Land Office, personal communication, 2014.

2. Williams and others, "Temperature as a potent driver of regional forest drought stress and tree mortality."

3. US Forest Service, Herger-Feinstein Quincy Library Group Forest Recovery Act Pilot Project, Status Report to Congress. http://www.fs.fed.us/r5/hfqlg/publica tions/congress_report/2010/FY10-HFQLG-Report-Final.pdf.

4. Jamieson, "Thinning project shows dramatic success," *Missoulian,* September 20, 2005.

5. Fiedler, Becker, and Haglund, "Preliminary guidelines for uneven-aged management in ponderosa pine"; Arno and others, "Restoring fire-dependent ponderosa pine forests in western Montana"; Fiedler, "Restoration treatments promote growth and reduce mortality of old-growth ponderosa pine (Montana)."

6. Covington and others, "Restoring ecosystem health in ponderosa pine forests of the Southwest."

7. Waltz and Covington, "Butterfly richness and abundance increase in restored ponderosa pine ecosystem"; Fulé and others, "Potential fire behavior is reduced following forest restoration treatments"; Laughlin and others, "Assessing targets for the restoration of herbaceous vegetation in ponderosa pine forests."

8. Zack and others, "A prospectus on restoring late successional forest structure to eastside pine ecosystems through large-scale, interdisciplinary research."

9. Nahban, foreword for *Ecological Restoration of Southwestern Ponderosa Pine Forests.*

10. J. Habeck, University of Montana, personal communication, 2014.

11. Cooper, "Changes in vegetation, structure, and growth of southwestern pine forests since white settlement"; Arno, Smith, and Krebs, "Old growth ponderosa pine and western larch stand structures: Influences of pre-1900 fires and fire exclusion"; Moore and others, "Comparison of historical and contemporary forest structure and composition on permanent plots in southwestern ponderosa pine forests"; Youngblood, Max, and Coe, "Stand structure in east-side old-growth ponderosa pine forests of Oregon and northern California."

12. Fiedler, Metlen, and Dodson, "Restoration treatment effects on stand structure, tree growth, and fire hazard in a ponderosa pine/Douglas-fir forest in Montana."

13. Fulé and others, "Measuring forest restoration effectiveness in reducing hazardous fuels."

14. Northwest Fire Science Consortium, "Influences on wildfire burn severity."

15. Fulé and others, "Comparing ecological restoration alternatives: Grand Canyon, Arizona"; Sala and Callaway, *Physiological Responses of Old Growth Ponderosa Pine and Western Larch to Restoration Cutting and Burning Treatments*; Fiedler and others, "Managing for old growth in frequent-fire landscapes."

16. US Forest Service, *Protecting People and Sustaining Resources in Fire-Adapted Ecosystems: A Cohesive Strategy*.

17. J. T. Williams, personal communication, 2014.

18. *Economist*, "Natural disasters."

19. M. Williams, Kaibab National Forest, personal communication, 2013.

20. Trachtman, "Fire fight."

21. Ibid.

22. J. A. Youtz, US Forest Service Southwestern Regional Office, personal communication, 2014.

23. Trachtman, "Fire fight."

24. Tholen, "Is collaboration working?"

25. H. Provencio, Coconino National Forest, personal communication, 2013.

26. Ibid.

27. W. Noble, Coconino National Forest, personal communication, 2013.

28. Fulé, Covington, and Moore, "Determining reference conditions for ecosystem management of southwestern ponderosa pine forests."

29. N. McCusker, Coconino National Forest, personal communication, 2013.

30. M. Lata, Coconino National Forest, personal communication, 2013.

31. Ibid.

32. J. Banks, Kaibab National Forest, personal communication, 2013.

33. H. Provencio, Coconino National Forest, personal communication, 2013.

34. Williams, *The Forest Service: Fighting for Public Lands*.

35. Phillips, *A Review of the Lakeview Federal Sustained Yield Unit, Fremont National Forest*.

36. Hanscom, "A timber town learns to care for the forest."

37. P. Harlan, Collins Pine Company (Lakeview, Oregon), personal communication, 2013.

38. Hanscom, "A timber town learns to care for the forest."

39. Ibid.

40. M. Anderson, the Wilderness Society (Portland, Oregon), personal communication, 2013.

41. Brown, "Getting from "no" to "yes": A conservationist's perspective."

42. C. Bienz, the Nature Conservancy (Klamath Falls, Oregon), personal communication, 2013.

43. A. McAdams, Lakeview Ranger District, personal communication, 2013.

44. C. Bienz, the Nature Conservancy (Klamath Falls, Oregon), personal communication, 2013.

45. Stephens and others, "Fire treatment effects on vegetation structure, fuels, and potential fire severity in western US forests."

Chapter 12. Protecting a Home and Its Forest

1. Cohen, *Examination of the Home Destruction in Los Alamos Associated with the Cerro Grande Fire.*
2. www.washingtondnr.wordpress.com/2010/09/21.
3. Mallon, "A story from the semi-arid eastside."
4. Forest and Rangelands, Success Stories, Landowner creates defensible space, http://www.forestsandrangelands.gov/success/stories/2009/nfp_2009_nm_fs_linf_fuels.shtml.
5. Johnson, "Governor: Homeowners must prepare for fire," *Missoulian*, June 23, 2009.
6. Pacific Northwest Extension, *Reducing Fire Risk on Your Forest Property.*
7. For instance, Arizona Wildfire Prevention and Information Website at www.wildlandfire.az.gov; and Pacific Northwest Extension guide, *Reducing Fire Risk on Your Forest Property*, available at www.ext.wsu.edu/forestry/documents/.
8. Mississippi State University Extension Service, *Prescribed Burning in Southern Pine Forests.*

Part II
Ponderosa Pines On and Off the Beaten Path

1. Muir, John, Letter to the author of *The Silva of North America*, June 7, 1898.
2. Summerhayes, *Vanished Arizona: Recollections of My Army Life*, p. 65.
3. C. Farris, US Geological Survey (Klamath Falls, Oregon), personal communication, 2014.
4. J. Malusa, University of Arizona, personal communication, 2014.
5. Bigelow, *On the Bloody Trail of Geronimo*, p. 202; C. Baisan, University of Arizona, personal communication.
6. Roberts, "The decline of Crystal Lake," *Los Angeles Times*, October 24, 1990.
7. Duffield, "The pines of Eddy Arboretum."
8. Forest History Society, Inventory of the William B. Laughead Papers, http://www.foresthistory.org/ead/Laughead_William_B.html.
9. US Geological Survey, *Geohydrology of Big Bear Valley, California.*
10. US Bureau of Land Management, *Pinelands Research Natural Area Management Plan.*
11. Kaelin, "Ute culturally scarred trees."
12. L. Swisher, San Juan National Forest, personal communication, 2014.
13. McClellan, *Timber: The Story of McPhee, Largest Lumbering Camp in Colorado.*
14. P. Brown, Rocky Mountain Tree-Ring Research (Fort Collins, Colorado), personal communication, 2014.
15. C. Hood, Idaho City Ranger District, personal communication, 2014.
16. US Forest Service, Bitterroot National Forest, History and Culture, Historic Alta Ranger Station, http://www.fs.usda.gov/detailfull/bitterroot/learning/history-culture/?cid=STELPRDB5160446.
17. D. J. Sandbak, Custer National Forest, personal communication, 2014.

18. Smalley, *History of the Northern Pacific Railroad*, p. 342.

19. Backus, "Primm for all time," *Missoulian*, July 7, 2005.

20. R. Means, Bureau of Land Management (Cheyenne, Wyoming), personal communication, 2014.

21. E. Guiberson, Bureau of Land Management (Dillon, Montana), personal communication, 2014.

22. S. Berndt (Chadron, Nebraska), personal communication, 2014.

23. Gardner, "Constructing a technological forest: Nature, culture, and tree-planting in the Nebraska Sand Hills."

24. R. Gilbert, Bessey Ranger District, personal communication, 2014.

25. Gardner, "Constructing a technological forest."

26. D. Charlet, College of Southern Nevada, personal communication, 2014.

27. Stumpff, *The People's Forest: Emerging Strategies on the Mescalero Apache Forest Reserves*.

28. Grissino-Mayer, Swetnam, and Adams, "The rare, old-aged conifers of El Malpais."

29. Lysne and McCormick, "Photo itinerary: Valles Caldera National Preserve."

30. Burke, "Keepers of the flame."

31. Allen, "A ponderosa pine natural area reveals its secrets."

32. Falk, "Scaling rules for fire regimes."

33. Potter and Green, "Ecology of ponderosa pine in western North Dakota."

34. Little, *Atlas of United States Trees*.

35. M. J. Roberts, Black Mesa, personal communication, 2014.

36. Richard, "A big tree monster at LaPine State Park," the *Oregonian*, June 5, 2008.

37. Fattig, "Tallest of the tall," *Mail Tribune* (Medford, Oregon), January 23, 2011.

38. Berry, "An ecological study of a disjunct ponderosa pine forest in the northern Great Basin in Oregon."

39. South Dakota Game, Fish, and Parks, Scenic Drives, Needles Highway, http://gfp.sd .gov/stateparks/directory/custer/activities/drives.aspx.

40. *Into the Green*, blog on Wordpress, http://intothegreen.wordpress.com/2010/04/15/.

41. E. Flores, Guadalupe Mountain National Park, personal communication.

42. Utah State University Extension, Forestry, the Potter Diaries, http://forestry.usu.edu /htm/ruralforests/forest-history/the-potter-diaries/.

43. McAvoy, *Utah Forest Landowner Education Program Newsletter*.

44. Utah History to Go, The Myths and Legends of Butch Cassidy, http://historytogo. utah.gov/utah_chapters/statehood_and_the_progressive_era/themythsandleg endsofbutchcassidy.html.

45. D. Page, Bureau of Land Management (Cedar City, Utah), personal communication, 2014.

46. Foster, "Westside story: restoration of a ponderosa pine forest at Fort Lewis Army Base."

47. US Fish and Wildlife Service, Turnbull National Wildlife Refuge pamphlet.

48. E. K. Dodson, Oregon State University, personal communication, 2013.

49. Vlahos, "The American icons."

50. R. Means, Bureau of Land Management (Cheyenne, Wyoming), personal communication, 2014.

51. Potter and others, "Haplotype distribution patterns in *Pinus ponderosa* (Pinaceae)."

52. R. Means, Bureau of Land Management (Cheyenne, Wyoming), personal communication, 2014.

53. Ibid.

54. Ibid.

55. Bridge River–Lillooet Country open community archive, http://www.cayoosh.net/hangtree.html

56. Ibid.

57. Turner and others, *Thompson Ethnobotany*.

58. R. Gray, R. W. Gray Consulting Ltd. (Chilliwack, BC), personal communication, 2015.

59. Rocky Mountain Trench Ecosystem Restoration Program, http://www.trench-er.com.

BIBLIOGRAPHY

Abarr, J. 1997. Inscription Rock. *Albuquerque Journal*, May 11.

Agee, J. K. 2002. The fallacy of passive management of western forest reserves. *Conservation Biology in Practice* 3: 18–25.

Allen, C. D. 1998. A ponderosa pine natural area reveals its secrets. In *Status and Trends of the Nation's Biological Resources*, edited by M. J. Mac, P. A. Opler, C. E. Puckett Haecker, and P. D. Doran, p. 551–52, US Geological Survey.

Anaconda Forest Products Company Records. 1890–1971. Maureen and Mike Mansfield Library Archives and Special Collections. University of Montana.

Anderson, I. V. 1933. *Application of Selective Logging to a Ponderosa Pine Operation in Western Montana*. University of Montana Paper 339.

Anderson, M. K. 2005. *Tending the Wild: Native American Knowledge and the Management of California's Natural Resources*. University of California Press.

Anderton, L., D. McAvoy, and M. Kuhns. *Native American Uses of Utah Forest Trees*. Utah Forest Facts NR/FF/018, Utah State University Cooperative Extension.

Arno, S. F. 1986. *Timberline: Arctic and Alpine Forest Frontiers*. Mountaineers Books.

Arno, S. F., and S. Allison-Bunnell. 2002. *Flames in Our Forest: Disaster or Renewal?* Island Press.

Arno, S. F., and C. E. Fiedler. 2005. *Mimicking Nature's Fire: Restoring Fire-Prone Forests in the West*. Island Press.

Arno, S. F., M. G. Harrington, C. E. Fiedler, and C. E. Carlson. 1995. Restoring fire-dependent ponderosa pine forests in western Montana. *Restoration and Management Notes* 13: 32–36.

Arno, S. F., H. Y. Smith, and M. A. Krebs. 1997. Old growth ponderosa pine and western larch stand structures: Influences of pre-1900 fires and fire exclusion. US Forest Service, Intermountain Research Station, Research Paper 495.

Backus, P. 2005. Primm for all time: Five Valleys Land Trust and Plum Creek Timber join forces to preserve old growth ponderosa. *Missoulian*, July 7.

Backus, P. 2009. Bitterroot National Forest: Officials seek options for thinning. *Missoulian*, July 8.

Baier, L. E. 2012. Reforming the Equal Access to Justice Act. *Journal of Legislation* 38(1): 1–70.

Baker, R. D., R. S. Maxwell, V. H. Treat, and H. C. Dethloff. 1988. *Timeless Heritage: A History of the Forest Service in the Southwest*. US Forest Service FS-409.

Ball, J. 2005. Forest health conditions and forest management practices on the Black Hills National Forest. Testimony provided 31 August, 2005, Hill City, SD.

Barrett, L. 1935. *A Record of Forest and Field Fires in California.* US Forest Service.

Bates, Carlos G. 1924. *Forest Types of the Central Rocky Mountains as Affected by Climate and Soil.* US Forest Service, Bulletin 1233.

Berry, R. W. 1963. An ecological study of a disjunct ponderosa pine forest in the northern Great Basin in Oregon. PhD diss., Oregon State University.

Bigelow, J. 1968. *On the Bloody Trail of Geronimo.* Westernlore Press.

Biondi, F. 1996. Decadal-scale dynamics at the Gus Pearson Natural Area: Evidence for inverse (a)symmetric competition? *Canadian Journal of Forest Research* 26: 1397–1406.

Biswell, H. 1974. Effects of fire on chaparral. In *Fire and Ecosystems,* edited by T. T. Kozlowski and C. Ahlgren, p. 321–64. Academic Press.

Biswell, H., H. H. Kallander, R. Komarek, R. Vogl, and H. Weaver. 1973. *Ponderosa Fire Management.* Tall Timbers Research Station, Misc. Publ. 2.

Bolle, A. W., R. W. Behan, G. Browder, T. Payne, W. L. Pengelly, R. E. Shannon, and R. F. Wambach, Select Committee of the University of Montana. 1970. *A University View of the Forest Service.* US Senate, Committee on Interior and Insular Affairs. Senate Document 91-115. Government Printing Office.

Botkin, D. 1990. *Discordant Harmonies: A New Ecology for the Twenty-First Century.* Oxford University Press.

Boyd, R. (editor). 1999. *Indians, Fire, and the Land in the Pacific Northwest.* Oregon State University Press.

Brown, P. M., M. R. Kaufman, and W. D. Shepperd. 1999. Long-term, landscape patterns of past fire events in a montane ponderosa pine forest of central Colorado. *Landscape Ecology* 14: 513–32.

Brown, P. M., C. L. Wienk, and A. J. Symstead. 2008. Fire and forest history at Mount Rushmore. *Ecological Applications* 18: 1984–99.

Brown, R. 2009. Getting from "no" to "yes": A conservationist's perspective. In *Old growth in a New World: A Pacific Northwest Icon Reexamined,* edited by T. A. Spies and S. L. Duncan, p. 149–57. Island Press.

Brown, T. J., B. L. Hall, and A. L. Westerling. 2004. The impact of twenty-first-century climate change on wildland fire danger in the western United States: An applications perspective. *Climatic Change* 62: 365–88.

Burk, D. A. 1970. *Clearcut Crisis: Controversy in the Bitterroot.* Jursnick Printing.

Burke, A. 2004. Keepers of the flame. *High Country News,* November 9.

Burns, R. L., and B. H. Honkala (tech. coordinators). 1990. *Silvics of North America,* vol.1, US Forest Service, Agriculture Handbook 654.

Callaham, R. Z. 2013. Pinus ponderosa*: A Taxonomic Review with Five Subspecies in the United States.* US Forest Service Research Paper PSW-RP-264.

Calloway, Colin G. 2003. *One Vast Winter Count: The Native American West before Lewis and Clark.* University of Nebraska Press.

Carle, D. 2002. *Burning Questions: America's Fight with Nature's Fire.* Praeger.

Carlson, C., S. Dobrowski, and H. D. Safford. 2009. *Angora Fire Vegetation Monitoring Annual Progress Report.* College of Forestry and Conservation, University of Montana.

Carson, R. 1962. *Silent Spring.* Houghton Mifflin.

Chapman, H. H. 1926. *Factors Determining Natural Reproduction of Longleaf Pine on Cut-Over Lands in LaSalle Parish, Louisiana.* Yale School of Forestry Bulletin 16.

Clary, D. A. 1986. *Timber and the Forest Service.* University Press of Kansas.

Coe, U. C. 1939. *Frontier Doctor.* MacMillan.

Cohen, J. D. 2000. *Examination of the Home Destruction in Los Alamos Associated with the Cerro Grande Fire.* US Forest Service, Rocky Mountain Research Station, Fire Sciences Lab, Fire Behavior Unit, Office Report, Missoula, MT.

Cooper, C. F. 1960. Changes in vegetation, structure, and growth of southwestern pine forests since white settlement. *Ecological Monographs* 30: 129–64.

Cooper, J. G. 1860. Report on the botany of the route. In *Reports of Explorations and Surveys to Ascertain the Most Practical and Economical Route for a Railroad from the Mississippi River to the Pacific Ocean.* Issac I. Stevens, US House of Representatives: 36th Congress, 1st Session, Ex. Doc. 56.

Covey, C. (translator and editor). 1983. *Cabeza de Vaca's Adventures in the Unknown Interior of America.* University of New Mexico Press.

Covington, W. W., P. Z. Fulé, M. M. Moore, S. C. Hart, T. E. Kolb, J. N. Mast, S. S. Sackett, and M. R. Wagner. 1997. Restoring ecosystem health in ponderosa pine forests of the Southwest. *Journal of Forestry* 95: 23–29.

Crawford, M. 1897. *Journal of Medorem Crawford.* Edited by F. G. Young. California Digital Library.

Cunningham, L. T. 1965. Famous old sawmills of the Shingletown Ridge. *The Covered Wagon,* p. 15–20. Shasta Historical Society.

DeBonis, J. 2009. Twenty years of activism. *Forest Magazine,* summer.

DeLasaux, M., and C. Starrs. 2012. Quincy Library Group seeks reauthorization from Congress. *The Forestry Source,* March.

DeQuille, D. 1876. *History of the Big Bonanza.* American Publishing Co.

DeWald, L. E., and M. F. Mahalovich. 2008. Historical and contemporary lessons from ponderosa pine genetic studies at the Fort Valley Experimental Forest, Arizona. In *Fort Valley Experimental Forest—A Century of Research 1908–2008,* edited by S. D. Olberding and M. M. Moore, p. 150–55, US Forest Service RMRS-P-55.

Douglass, A. E. 1929. The secret of the Southwest solved by talkative tree rings. *National Geographic Magazine* 56: 736–70.

Duffield, J. W. 1949. The pines of Eddy Arboretum. *Arboretum Bulletin* XII (4). Available at http://www.fs.fed.us/psw/publications/duffield/psw_1949_duffield001.pdf.

Earley, L. S. 2004. *Looking for Longleaf: The Fall and Rise of an American Forest.* University of North Carolina Press.

Economist. 2012. Natural disasters: Counting the costs of calamities. *Economist* 402 (8767): 11, 60–62.

Egan, T. 2009. *The Big Burn.* Houghton Mifflin.

English, N. B., J. L. Betancourt, J. S. Dean, and J. Quade. 2001. Strontium isotopes reveal distant sources of architectural timbers in Chaco Canyon, New Mexico. *Proceedings of the National Academy of Sciences USA* 98: 11891–96.

Evans, J. W. 1991. *Powerful Rockey: The Blue Mountains and the Oregon Trail.* Pika Press.

Falk, D. A. 2004. Scaling rules for fire regimes. PhD diss., University of Arizona.

Fanselow, J. 2001. *Traveling the Oregon Trail*. Globe Pequot Press.

Faris, J. T. 1930. *Roaming the Rockies*. Farrar and Rinehart.

Fattig, P. 2011. Tallest of the tall. *Mail Tribune* (Medford, OR), January 23.

Fettig, C. J., K. D. Klepzig, R. F. Billings, A. S. Munson, T. E. Nebeker, J. F. Negrón, and J. T. Nowak. 2007. The effectiveness of vegetation management practices for prevention and control of bark beetle infestations in coniferous forests of the western and southern United States. *Forest Ecology and Management* 238: 24–53.

Fiedler, C. E. 2000. Restoration treatments promote growth and reduce mortality of old-growth ponderosa pine (Montana). *Ecological Restoration* 18: 117–19.

Fiedler, C. E., R. Becker, and S. Haglund. 1988. Preliminary guidelines for uneven-aged management in ponderosa pine. In *Ponderosa Pine: The Species and Its Management*, p. 235–41. Washington State University Cooperative Extension.

Fiedler, C. E., E. K. Dodson, and K. L. Metlen. 2013. Exotic plant response to forest disturbance in the western United States. In *Invasive Plant Ecology*, edited by S. Jose, H. P. Singh, D. R. Batish, and R. K. Kohli, p. 93–112. CRC Press.

Fiedler, C. E., P. Friederici, M. Petruncio, C. Denton, and W. D. Hacker. 2007. Managing for old growth in frequent-fire landscapes. *Ecology and Society* 12(2): 20.

Fiedler, C. E., K. L. Metlen, and E. K. Dodson. 2010. Restoration treatment effects on stand structure, tree growth, and fire hazard in a ponderosa pine/Douglas-fir forest in Montana. *Forest Science* 56: 18–31.

Foster, J. R. Westside story: Restoration of a ponderosa pine forest at Fort Lewis Army Base. In *Prairie Landowners Guide for Western Washington*. South Puget Sound Prairie Landscape Working Group, p. 217–30. Available at http://w.southsound prairies.org/tech/Poderosa%20Pine%20Restoration.pdf.

Franklin, J. F., D. R. Berg, A. B. Carey, and R. A. Hardt. 2006. Old-growth conifer forests. In *Restoring the Pacific Northwest: The Art and Science of Ecological Restoration in Cascadia*, edited by D. Apostol and M. Sinclair, p. 112–21. Island Press.

Frémont, J. C. 2001. *Memoirs of My Life*. Rowman & Littlefield.

Fulé, P. Z., W. W. Covington, and M. M. Moore. 1997. Determining reference conditions for ecosystem management of southwestern ponderosa pine forests. *Ecological Applications* 7: 895–908.

Fulé, P. Z., W. W. Covington, H. Smith, J. Springer, T. A. Heinlein, K. Huisinga, and M. Moore. 2002. Comparing ecological restoration alternatives: Grand Canyon, Arizona. *Forest Ecology and Management* 170: 19–41.

Fulé, P. Z., J. E. Crouse, , T. A. Heinlein, M. M. Moore, W. W. Covington, and G. Verkamp. 2003. Mixed-severity fire regime in a high-elevation forest of Grand Canyon, Arizona, USA. *Landscape Ecology* 18: 465–86.

Fulé, P. Z., C. McHugh, C. T. A. Heinlein, and W. W. Covington. 2001. Potential fire behavior is reduced following forest restoration treatments. In *Ponderosa Pine Ecosystems Restoration and Conservation: Steps Toward Stewardship*, compiled by R. K. Vance and others, p. 28–35, US Forest Service RMRS-P-22.

Fulé, P. Z., A. E. M. Waltz, W. W. Covington, and T. A. Heinlein. 2001. Measuring forest restoration effectiveness in reducing hazardous fuels. *Journal of Forestry* 99: 24–29.

Galloway, J. D. 1950. *The First Transcontinental Railroad: Central Pacific, Union Pacific*. Simmons-Boardman.

Gardner, R. 2009. Constructing a technological forest: Nature, culture, and tree-planting in the Nebraska Sand Hills. *Environmental History* 14: 275–97.

Garrett, L. D., M. H. Soulen, and J. R. Ellenwood. 1995. After 100 years of forest management: The North Kaibab. In *Proceedings of the Third Biennial Conference of Research on the Colorado Plateau*, National Park Service Transactions and Proceedings Series, edited by C. Van Riper and E. Deschler, p. 129–49.

Geyer, C. A. 1846. Notes on the vegetation and general character of the Missouri and Oregon Territories. *London Journal of Botany* 5: 204.

Graham, R. T., and T. B. Jain. 2005. *Proceedings of the Symposium on Ponderosa Pine*. US Forest Service, General Technical Report PSW-198.

Graves, H. S. 1899. The Black Hills Forest Reserve. In *The Nineteenth Annual Report of the Survey, 1897–98. Part V, Forest Reserves*, US Geological Survey, p. 67–164.

Graves, H. S. 1920. The torch in the timber. *Sunset* 44: 37–40.

Gray, A. A., F. P. Farquhar, and W. S. Lewis. 1930. Camels in western America. *Quarterly of the California Historical Society*, December.

Greene, S. W. 1931. The forest that fire made. *American Forests* 37 (October): 583.

Grissino-Mayer, H. D., T. W. Swetnam, and R. K. Adams. 1997. The rare, old-aged conifers of El Malpais—Their role in understanding climatic change in the American Southwest. *New Mexico Bureau of Mines and Mineral Resources Bulletin* 156: 155–61.

Gruell, G. E. 1983. *Fire and Vegetative Trends in the Northern Rockies: Interpretations from 1871–1982 Photographs*. US Forest Service, Intermountain Forest and Range Experimental Station, General Technical Report 158.

Gruell, G. E. 2001. *Fire in Sierra Nevada Forests: A Photographic Interpretation of Ecological Change Since 1849*. Mountain Press.

Haller, J. R., and N. R. Vivrette. 2011. Ponderosa pine revisited. *Aliso* 29: 53–57.

Halliday, W. R. 1969. The forgotten father of North Cascades National Park. *Seattle Times*, March 16.

Hammer, R. B., S. Stewart, and V. Radeloff. 2009. Demographic trends, the wildland-urban interface, and wildfire management. *Society and Natural Resources* 22(8): 777–82.

Hanscom, G. 2004. A timber town learns to care for the forest. *High Country News*, September 27.

Hart, J. 1981 The ethnobotany of the Northern Cheyenne Indians of Montana. *Journal of Ethnopharmacology* 4: 1–55.

Hart, J. 1992. *Montana Native Plants and Early Peoples*. Montana Historical Society Press.

Headwaters Economics. 2009. Solutions to the rising costs of fighting fires in the wildland-urban interface. Published online at www.headwaterseconomics.org.

Hillis, M. 2007. Tweak laws to keep fringe from environmental lawsuit frenzy. *Missoulian*, April 8.

Hirt, P. W. 1994. *A Conspiracy of Optimism: Management of the National Forests since World War Two*. University of Nebraska Press.

Hoffman, A. F. C., and T. Krueger. 1949. Forestry in the Black Hills. In *Trees* (Part 2), USDA Yearbook of Agriculture Series, p. 319–26. US Government Printing Office.

Holtmeier, F. 1973. Geoecological aspects of timberlines in northern and central Europe. *Arctic and Alpine Research* 5(3): A45–A54.

Hoover, M. B., H. E. Rensch, E. G. Rensch, and W. N. Abeloe. Revised by D. E. Kyle. 2002. *Historic Spots in California*. Stanford University Press.

Horsted, P. 2006. *The Black Hills Yesterday and Today*. Golden Valley Press.

Howe, G. 1995. Genetic effects of uneven-aged management. In *Uneven-Aged Management: Opportunities, Constraints and Methodologies*, Montana Forest and Conservation Experiment Station Miscellaneous Publication No. 56, edited by K. L. O'Hara, p. 27–32.

Hoxie, G. L. 1910. How fire helps forestry. *Sunset* 34: 145–51.

Hudson, M. 2011. *Fire Management in the American West: Forest Politics and the Rise of Megafires*. University Press of Colorado.

Hunt, R. H. 2003. The beginning of the Lewis and Clark Expedition: Jefferson's secret message to Congress. *Calendar Features*, January (online, National Archives.)

Hutchinson, W. H. 1983. *California Heritage: A History of Northern California Lumbering*. Duke University Press.

Iniguez, J. M., T. W. Swetnam, and C. H. Baisan. 2009. Spatially and temporally variable fire regime on Rincon Peak, Arizona, USA. *Fire Ecology* 5: 3–21.

Jackson, D. 1974. *Custer's Gold: The United States Cavalry Expedition of 1874*. University of Nebraska Press.

Jamieson, M. 2005. Thinning project shows dramatic success. *Missoulian*, September 20.

Johnson, B. 1978. *Chips and Sawdust*. The Press Room.

Johnson, C. 2009. Governor: Homeowners must prepare for fire. *Missoulian*, June 23.

Joslin, L. 2007. *Ponderosa Promise: A History of US Forest Service Research in Central Oregon*. US Forest Service, Pacific Northwest Research Station General Technical Report PNW-GTR-711.

Kaelin, C.R. 2003. Ute culturally scarred trees. http://www.pikespeakhsmuseum.org .

Kahn, E. M. 2006. Flume with a fiew: Saving an American engineering marvel. *Icon*, summer: 40–43.

Kilgore, B. M. and G. A. Curtis. 1987. *Guide to Understory Burning in Ponderosa Pine–Larch–Fir Forests in the Intermountain West*. US Forest Service, Intermountain Research Station, General Technical Report 233.

Kingsbury, L. A., and G. Dixon. 2012. American Indian culturally modified ponderosa pine trees on the Payette National Forest. Unpublished report on file at Payette National Forest Supervisor's office in McCall, Idaho.

Kitts, J. 1919. Preventing forest fires by burning litter. *Timberman* July.

Koch, E. 1935. The passing of the Lolo Trail. *Journal of Forestry* 33(2): 98–104.

Koch, E. 1998. *Forty Years a Forester, 1903–1943*. Mountain Press.

Krause, H., and G. Olson. 1974. *Prelude to Glory*. Brevet Press.

Lang, D. M., and S. S. Stewart. 1910. *Reconnaissance of the Kaibab National Forest. US Forest Service*. Unpublished report. On file at Kaibab National Forest Supervisor's Office, Williams, AZ.

Lanner, R. L. 1983. *Trees of the Great Basin*. University of Nevada Press.

Laughlin, D. C., M. M. Moore, J. D. Bakker, C. A. Casey, J. D. Springer, P. Z. Fulé, and W. W. Covington. 2006. Assessing targets for the restoration of herbaceous vegetation in ponderosa pine forests. *Restoration Ecology* 14: 548–60.

Leavengood, B. 2000. History of Phantom Ranch. *Grand Canyon Explorer* 4 (9): 1–3.

Leiberg, J. B. 1900. The Bitterroot Forest Reserve. *20th Annual Report, US Geological Survey, Part 5:* 317–410.

Leopold, A. 1949. *A Sand County Almanac and Sketches Here and There.* Oxford University Press.

Leopold, A., S. S. Cain, C. Cottam, I. Gabrielson, and T. Kimball. 1963. Wildlife management in the national parks. *Transactions of the North American Wildlife and Natural Resources Conference* 28: 28–45.

Lewis, E. J., contributor. 1880. *History of Victorian Tehama County, California.* Elliott & Moore.

Lister, R. H., and F. C. Lister. 1984. *Chaco Canyon: Archaeology and Archaeologists.* University of New Mexico Press.

Little, E. L., Jr. 1971. *Atlas of United States Trees: Volume 1. Conifers and Important Hardwoods.* US Forest Service Miscellaneous Publication 1146.

Loosle, B. 2004. Ponderosa bark used for food, glue, and healing. *Utah Forest News* 8 (3): 4–5.

Lutts, R. H. 1992. The trouble with Bambi: Walt Disney's Bambi and the American vision of nature. *Forest and Conservation History* 36: 160–71.

Lysne, J., and P. McCormick. Photo Itinerary: Valles Caldera National Preserve. *Nature Photographers Online Magazine.* http://www.naturephotographers.net /articles1105/pm1105-1.html.

Maclean, N. 1976. *A River Runs Through It and Other Stories.* University of Chicago Press.

Mahar, James Michael. 1953. *Ethnobotany of the Oregon Paiutes of the Warm Springs Indian Reservation.* BA thesis. Reed College.

Mallon, C. 2012. A story from the semi-arid eastside. *Northwest Woodlands* 28(3): 16–19.

Mann, Charles C. 2005. *1491: New Revelations of the Americas before Columbus.* Alfred A. Knopf.

Marshall, P. 2007. The high cost of fire. *Forest Magazine,* fall: 14–20.

Matthews, A. J. 2002. *Montana Main Streets: Vol. 6. A Guide to Historic Missoula.* Montana Historical Society Press.

McAvoy, D. 2010. *Utah Forest Landowner Education Program Newsletter* 14 (3). Utah State University Cooperative Extension.

McClellan, S. 1970. *Timber: The Story of McPhee, Largest Lumbering Camp in Colorado.* Dolores Star Press.

McKay, K. L. 1994. *Trails of the Past: Historical Overview of the Flathead National Forest, 1800–1960.* Final report of a historic overview prepared under agreement with the Flathead National Forest.

Meyer, W. H. 1934. *Growth in Selectively Cut Ponderosa Pine Forests of the Pacific Northwest.* USDA Technical Bulletin 407.

Mills, E. A. 1900. *The Story of a Thousand-Year Pine.* Houghton Mifflin.

Mississippi State University Extension Service. 2012. *Prescribed Burning in Southern Pine Forests: Fire Ecology, Techniques, and Uses for Wildlife Management*. Publication 2283.

Missoulian. Op-ed, March 19, 2007.

Montana Record-Herald (Helena). 1926. Giant yellow pine 1,100 years old is felled near Evaro. January 19.

Moore, M. M., D. W. Huffman, P. Z. Fulé, W. W. Covington, and J. E. Crouse. 2004. Comparison of historical and contemporary forest structure and composition on permanent plots in southwestern ponderosa pine forests. *Forest Science* 50: 162–76.

Morgan, M. 1955. *The Last Wilderness*. Viking Press.

Mortimer, M. J., and R. W. Malmsheimer. 2011. The Equal Access to Justice Act and US Forest Service land management: Incentives to litigate? *Journal of Forestry* 109: 352–58.

Moulton, G. E. (editor). 1988. *The Journals of the Lewis and Clark Expedition*, vol. 5, University of Nebraska Press.

Muir, J. 1907. *The Mountains of California*. The Century Co.

Munger, T. T. 1917. *Western Yellow Pine in Oregon*. USDA Department Bulletin 418.

Murphy, A. 1994. *Graced by Pines: The Ponderosa Pine in the American West*. Mountain Press.

Nahban, G. P. 2003. Foreword. In *Ecological Restoration of Southwestern Ponderosa Pine Forests*, edited by P. Friederici. Island Press.

Nash, S. E. 1999. *Time, Trees, and Prehistory: Tree-Ring Dating and the Development of North American Archaeology 1914–1950*. University of Utah Press.

Nelson, T. C. 1979. Fire management policy in the national forests: A new era. *Journal of Forestry* 77: 723–25.

Nijhuis, M. 2012. Forest fires: Burn out. *Nature* 489: 352–54.

Northwest Fire Science Consortium. 2014. Influence on wildfire burn severity: Treatment and landscape drivers in an extreme fire event. Research Brief 5. Oregon State University.

Odum, E. P. 1989. *Ecology and Our Endangered Life-Support Systems*. Sinauer Associates.

Olberding, S. D. 2008. *"It Was a Young Man's Life": G. A. Pearson*. US Forest Service RMRS-P-53CD. Rocky Mountain Research Station, Ogden, UT.

Olberding, S. D., D. Hueber, and C. Edminster. 2007. *Fort Valley Experimental Forest Historical Photographs*. Rocky Mountain Research Station, US Forest Service.

Pacific Northwest Extension. 2010. *Reducing Fire Risk on Your Forest Property*. PNW 618, www.ext.wsu.edu/forestry/documents/ pnw618.

Palladinos, L. B. 1884. *Anthony Ravalii, S.J.: Forty Years a Missionary in the Rocky Mountains*. California Digital Library.

Palmer, G. 1975. Shuswap Indian ethnobotany. *Syesis* 8: 29–51.

Paxon, J. 2007. *The Monster Reared His Ugly Head: The Story of the Rodeo-Chediski Fire and Fire as a Tool of Nature*. Cedarhill Publishing.

Pearson, G. A. 1910. *Reproduction of Western Yellow Pine in the Southwest*. US Forest Service Circular 174.

Pearson, G. A. 1923. *Natural Reproduction of Western Yellow Pine in the Southwest.* USDA Department Bulletin 1105.

Pearson, G. A. 1950. *Management of Ponderosa Pine in the Southwest.* USDA Agriculture Monograph No. 6.

Peattie, D. C. 1950. *A Natural History of Western Trees.* Houghton Mifflin.

Peterson, J. 2011. Good policy and good intentions won't stop big wildfires. *High Country News,* October 17.

Phelps, W. L. 1939. *Autobiography with Letters.* Oxford University Press.

Phillips, R. 2010. *A Review of the Lakeview Federal Sustained Yield Unit, Fremont National Forest, 2000–2009.* Phillips Economic Solutions.

Pierce, J. L., G. A. Meyer, and A. J. T. Jull. 2004. Fire-induced erosion and millennial-scale climate change in northern ponderosa pine forests. *Nature* 432: 87–90.

Pinchot, G. 1899. The relation of forests and forest fires. *National Geographic* 10: 393–403.

Pinchot, G. 1998. *Breaking New Ground.* Island Press.

Potter, K. M., V. D. Hipkins, M. F. Mahalovich, and R. E. Means. 2013. Haplotype distribution patterns in *Pinus ponderosa* (Pinaceae): Range-wide evolutionary history and implications for conservation. *American Journal of Botany* 100: 1562–79.

Potter, L. D., and D. L. Green. 1964. Ecology of ponderosa pine in western North Dakota. *Ecology* 45: 10–23.

Powell, J. W. 1891. *Testimony to Congress: Eleventh annual report of the US Geological Survey, 1889–1890. Part 2: Irrigation.*

Powell, J. W. 1961. *The Exploration of the Colorado River and its Canyons.* Dover Publications Inc.

Prichard, S. J., and M. C. Kennedy. 2014. Fuel treatments and landform modify landscape patterns of burn severity in an extreme fire event. *Ecological Applications* 24(3): 571–90.

Pyne, S. J. 1982. *Fire in America: A Cultural History of Wildland and Rural Fire.* Princeton University Press.

Pyne, S. J. 2001. *Year of the Fires: The Story of the Great Fires of 1910.* Viking Penguin.

Radeloff, V. C., and others. 2010. Housing growth in and near United States protected areas limits their conservation value. *Proceedings of the National Academy of Sciences* 107(2): 940–45.

Reed, P. F. 2004. *The Puebloan Society of Chaco Canyon.* Greenwood Publishing Group.

Reynolds, A. C., J. L Betancourt, J. Quade, P. J. Patchett, J. S. Dean and J. Stein. 2005. [87]Sr/[86]Sr sourcing of ponderosa pine used in Anasazi great house construction at Chaco Canyon, New Mexico. *Journal of Archaeological Science* 32: 1061–75.

Richard, T. 2008. A big tree monster at LaPine State Park. *The Oregonian* June 5.

Roberts, R. 1990. The decline of Crystal Lake. *Los Angeles Times,* October 24.

Robertson, F. D. 1992. *Ecosystem Management of the National Forests and Grasslands.* Memo to Regional Foresters and Research Station Directors, US Forest Service, June 4.

Rogers, P. 1996. *Disturbance Ecology and Forest Management: A Review of the Literature*. US Forest Service, Intermountain Research Station, General Technical Report 336.

Sala, A., and R. Callaway. 2004. *Physiological Responses of Old Growth Ponderosa Pine and Western Larch to Restoration Cutting and Burning Treatments*. Final Report, US Forest Service, Rocky Mountain Research Station Research Joint Venture Agreement 99563, Fire Sciences Lab, Missoula, MT.

Salish–Pend d'Oreille Culture Committee and Elders Cultural Advisory Council. 2005. *The Salish People and the Lewis and Clark Expedition*. University of Nebraska Press.

Sargent, C. S. 1897. *The Silva of North America: Vol. XI. Coniferae (Pinus)*. Houghton Mifflin.

Schiff, A. 1962. *Fire and Water: Scientific Heresy in the Forest Service*. Harvard University Press.

Schubert, G. H. 1974. *Silviculture of Southwestern Ponderosa Pine: The Status of Our Knowledge*. US Forest Service, Rocky Mountain Forest and Range Experiment Station, Research Paper RM-123.

Sherriff, R. L., and T. T. Veblen. 2007. A spatially-explicit reconstruction of historical fire occurrence in the ponderosa pine zone of the Colorado Front Range. *Ecosystems* 9: 1342–47.

Show, S. B., and E. I. Kotok. 1924. *The Role of Fire in the California Pine Forests*. USDA Department Bulletin 1294.

Shuford, B. 1957. Logging in Shasta County. In *The Covered Wagon*. Shasta Historical Society, p. 34–36.

Simpson, J. H. 2003. *Navaho Expedition: Journal of a Military Reconnaissance from Santa Fe, New Mexico to the Navaho Country, Made in 1849*. Edited by F. McNitt. University of Oklahoma Press.

Smalley, E. V. 1883. *History of the Northern Pacific Railroad*. G. P. Putnam's Sons.

Smith, D. 1992. The historic Blue Ridge Flume of Shasta and Tehama Counties, California. Paper 2. In *Gold and Lumber: Two Papers on Northern California History and Archaeology*. Bureau of Land Management.

Smith, H. Y., and S. Arno. 1999. *Eighty-Eight Years of Change in a Managed Ponderosa Pine Forest*. US Forest Service, Rocky Mountain Research Station, General Technical Report 23.

Snider, G., P. J. Daugherty, and D. Wood. 2006. The irrationality of continued fire suppression. *Journal of Forestry* 104: 431–37.

South Dakota State Historical Society and South Dakota Department of History. 1914. *South Dakota Historical Collections*, vol. 7. State Publishing Company.

Speer, J. H. 2010. *Fundamentals of Tree Ring Research*. University of Arizona Press.

Spencer, B. G. 1956. *The Big Blow-Up*. Caxton Printers.

Spokesman-Review (Spokane). 1991. Op-ed, "Hangman Hills [fire] lessons largely ignored," October 20.

Standley, P. C. 1912. Some useful native plants of New Mexico. In *Smithsonian Institute Annual Report of 1911*, Smithsonian Institution, p. 447–62.

Steen, H. K. 1976. *The US Forest Service: A History.* University of Washington Press.

Stephens, S. L., J. J. Moghaddas, C. Edminster, C. E. Fiedler, S. Hasse, M. Harrington, J. E. Keeley, J. D. McIver, K. Metlen, C. N. Skinner, and A. Youngblood. 2009. Fire treatment effects on vegetation structure, fuels, and potential fire severity in western U.S. forests. *Ecological Applications* 19: 305–20.

Stewart, O. C. 2002. *Forgotten Fires: Native Americans and the Transient Wilderness.* University of Oklahoma Press.

Stoddard, H. L. 1931. *The Bobwhite Quail: Its Habits, Preservation, and Increase,* Charles Scribner's Sons.

St. Paul Pioneer. 1874. Newspaper report, July 3.

Stumpff, L. M. *The People's Forest: Emerging Strategies on the Mescalero Apache Forest Reserves.* Evergreen State College. Available at http://nativecases.evergreen.edu/docs/stumpff_mescalero_forest.pdf.

Summerhayes, M. 1908. *Vanished Arizona: Recollections of My Army Life.* J. B. Lippincott.

Swanson, F. H. 2011. *The Bitterroot and Mr. Brandborg: Clearcutting and the Struggle for Sustainable Forestry in the Northern Rockies.* University of Utah Press.

Swetnam, T. W. 1984. Peeled ponderosa pine trees: A record of inner bark utilization by Native Americans. *Journal of Ethnobiology* 4: 177–90.

Swetnam, T. W. 1993. Fire history and climate change in giant sequoia groves. *Science* 262: 885–89.

Swetnam, T. W., and C. H. Baisan. 1996. Historical fire regime patterns in the southwestern United States since 1700. In *Fire Effects in Southwestern Forests,* Proceedings of the Second La Mesa Fire Symposium, March 29–31, 1994, Los Alamos, NM. US Forest Service, General Technical Report RM-GTR-286, edited by C. D. Allen, p. 11–32.

Symstad, A. J., and M. Bynum. 2005. *The Extent and Significance of Old-Growth Ponderosa Pine Forest at Mount Rushmore National Memorial.* Report to Mount Rushmore National Memorial. National Park Service, Keystone, South Dakota.

Tehama County, California: Illustrations Descriptive of Its Scenery, Fine Residences, Public Buildings, Manufactories, Hotels, Farm Scenes, Business Houses, Schools, Churches, Mines, Mills, Etc. Reprinted in 1975 by California History Books.

Tholen, R. 2013. Is collaboration working? Another reader's opinion. *Evergreen Magazine.* Spring.

Thomas, J. W. 2002. Dynamic vs. static management in a fire-influenced landscape—the Northwest Forest Plan. Text of presentation at the conference *Fire in Oregon Forests,* Oregon Forest Resources Institute, Portland.

Tidwell, T. 2011. Statement before the Committee on Energy and Natural Resources, Subcommittee on Public Lands and Forests, US Senate. August 3.

Toole, K. Ross. 1959. *Montana: An Uncommon Land.* University of Oklahoma Press.

Trachtman, P. 2003. Fire fight. *Smithsonian* 34: 42–52.

Turner, N. J., L. C. Thompson, M. T. Thompson, and A. Z. York. 1990. *Thompson Ethnobotany: Knowledge and Usage of Plants by the Thompson Indians of British Columbia.* Memoir No. 3. Royal British Columbia Museum.

Urbaniak, R. 2007. *Anasazi of Southwest Utah: The Dance of Shadow and Light.* Natural Frequency with Sanctuary House Press.

US Bureau of Land Management. 1987. *Pinelands Research Natural Area Management Plan*. Susanville District.

US Congress. *Multiple-Use Sustained Yield Act of 1960*. 86th Congress. Public Law 86-517.

US Fish and Wildlife Service. 2011. *Revised Recovery Plan for the Northern Spotted Owl* (Strix occidentalis caurina).

US Fish and Wildlife Service. Turnbull National Wildlife Refuge pamphlet.

US Forest Service. 2000. *Management Response to the General Accounting Office Report GAO/RCED-99-65, Protecting People and Sustaining Resources in Fire-Adapted Ecosystems: A Cohesive Strategy*. Unpublished report.

US Geological Survey. 2012. *Geohydrology of Big Bear Valley, California, Phase I: Geologic Framework, Recharge, and Preliminary Assessment of the Source and Age of Groundwater*. Scientific Investigations Report 2012-5100.

Vander Wall, S. B., and R. P. Balda. 1977. Coadaptations of the Clark's nutcracker and the piñon pine for efficient seed harvest and dispersal. *Ecological Monographs* 47: 89–111.

Van Pelt, R. 2001. *Forest Giants of the Pacific Coast*. University of Washington Press.

Vestal, P. A. 1952. The ethnobotany of the Ramah Navaho. *Papers of the Peabody Museum of American Archaeology and Ethnology* 40: 1–94.

Vlahos, J. 2004. The American icons: Twelve classic trips in the landscapes of legend. *National Geographic Adventure Magazine*, April.

Vosick, D. 2012. *An Era of Megafires*. Apache-Sitgreaves National Forests.

Wallace, E. S. 1955. *The Great Reconnaissance: Soldiers, Artists, and Scientists on the Frontier 1848–1861*. Little, Brown and Company.

Waltz, A. E. M., and W. W. Covington. 1999. Butterfly richness and abundance increase in restored ponderosa pine ecosystem (Arizona*). Ecological Restoration* 17: 244–46.

Weaver, H. 1943. Fire as an ecological and silvicultural factor in the ponderosa pine region of the Pacific Slope. *Journal of Forestry* 41: 7–14.

Weaver, H. 1968. Fire and its relationship to ponderosa pine. *Proceedings — Tall Timbers Fire Ecology Conference* 7: 128.

Weekly Sentinel (Red Bluff, California). 1874. Article from August 1.

Wellner, C. A. 1976. *Frontiers of Forestry Research — Priest River Experimental Forest, 1911–1976*. US Forest Service, Intermountain Forest and Range Experiment Station.

Westerling, A. L., H. G. Hidalgo, D. R. Cayan, and T. W. Swetnam. 2006. Warming and earlier spring increase western US forest wildfire activity. *Science* 313: 940–43.

Western Forestry Leadership Coalition. 2010. *The True Cost of Wildfire in the Western US*. Western Forestry Leadership Coalition.

Whipple, A. W. 1856. *Reports of Explorations and Surveys to Ascertain the Most Practicable and Economical Route for a Railroad from the Mississippi River to the Pacific Ocean. Vol. III: Route near the 35th parallel*. Beverly Tucker Printer.

White, S. E. 1920. Woodsmen, spare those trees! *Sunset* 44: 22–26.

White, T. 1954. *Scarred Trees in Western Montana*. Montana State University Anthropology and Sociology Papers, No. 17.

Willey, N. B. 1881. *Nez Perce News* (Lewiston, Idaho Territory) June 9.

Williams, A. P., C. D. Allen, A. K. Macalady, D. Griffin, C. A. Woodhouse, D. M. Meko, T. W. Swetnam, S. A. Rauscher, R. Seager, H. D. Grissino-Mayer, J. S. Dean, E. R. Cook, C. Gangodagamage, M. Cai, and N. G. McDowell. 2012. Temperature as a potent driver of regional forest drought stress and tree mortality. *Nature Climate Change* 3: 292–97.

Williams, G. W. 2007. *The Forest Service: Fighting for Public Lands*. Greenwood Press.

Winkler, C. 1976. Typed transcript of an oral history interview with Charles Winkler, interviewed by Joe Bennett and Doug Jones, April 27, 1976, Idaho State Historical Society.

Winship, G. P. 1896. "The Coronado Expedition, 1540–1542." Castañeda report in Spanish and accompanying English translation, In *Fourteenth Annual Report of the Bureau of Ethnology of the Smithsonian Institution, 1892–93*, Part I, p. 339–615. Smithsonian Institution.

Woolsey, T. S. 1911. *Western Yellow Pine in Arizona and New Mexico*. US Forest Service Bulletin 101.

Worster, D. 1994. *Nature's Economy: A History of Ecological Ideas*. Cambridge University Press.

Worster, D., S. Armitage, M. P. Malone, D. J. Weber, and P. N. Limerick. 1989. *The Legacy of Conquest* by Patricia Nelson Limerick: A panel of appraisal. *Western Historical Quarterly* 20 (3): 303–22.

Yankton Press. Articles from March 13 and March 20, 1872.

Youngblood, A., T. Max, and K. Coe. 2004. Stand structure in east-side old-growth ponderosa pine forests of Oregon and northern California. *Forest Ecology and Management* 199: 191–217.

Zack, S., W. F. Laudenslayer, T. L. George, C. Skinner, and W. Oliver. 1999. A prospectus on restoring late successional forest structure to eastside pine ecosystems through large-scale, interdisciplinary research. In *First Biennial North American Forest Ecology Workshop*, p. 343–55, Society of American Foresters.

INDEX

Page numbers in bold face refer to photos or figures.

249

ABOUT THE AUTHORS

Carl E. Fiedler earned a PhD in forestry and ecology from the University of Minnesota. He retired in 2007 after stints in the US Army, US Forest Service Intermountain Research Station, and twenty-five years as professor of forest management at the University of Montana. During summers, he presented short courses on ponderosa pine management, fire, and restoration forestry throughout the West. For him, nothing beats fishing with his brother Dave on one of Montana's many legendary trout streams.

Stephen F. Arno obtained a PhD in forestry and plant science from the University of Montana in 1970. He was a research forester with the USDA Fire Sciences Laboratory in Missoula, Montana, before retiring in 1999. He has practiced restoration forestry on his family's ponderosa pine forest for more than forty years and has written several books related to trees and forests. A committed denizen of the forest, he enjoys the long process of harvesting firewood and using it to heat the family home.